Driver Distraction

Transportation Human Factors: Aerospace, Aviation, Maritime, Rail, and Road Series

Series Editor:
Professor Neville A. Stanton, University of Southampton, UK

Automobile Automation
Distributed Cognition on the Road
Victoria A. Banks and Neville A. Stanton

Eco-Driving
From Strategies to Interfaces
Rich C. McIlroy and Neville A. Stanton

Driver Reactions to Automated Vehicles
A Practical Guide for Design and Evaluation
Alexander Eriksson and Neville A. Stanton

Systems Thinking in Practice
Applications of the Event Analysis of Systemic Teamwork Method
Paul Salmon, Neville A. Stanton, and Guy Walker

Individual Latent Error Detection (I-LED)
Making Systems Safer
Justin R.E. Saward and Neville A. Stanton

Driver Distraction
A Sociotechnical Systems Approach
Katie J. Parnell, Neville A. Stanton, and Katherine L. Plant

For more information about this series, please visit: https://www.crcpress.com/ Transportation-Human-Factors/book-series/CRCTRNHUMFACAER

Driver Distraction
A Sociotechnical Systems Approach

Katie J. Parnell, Neville A. Stanton and
Katherine L. Plant

CRC Press is an imprint of the
Taylor & Francis Group, an **informa** business

CRC Press
Taylor & Francis Group
6000 Broken Sound Parkway NW, Suite 300
Boca Raton, FL 33487-2742

© 2019 by Taylor & Francis Group, LLC
CRC Press is an imprint of Taylor & Francis Group, an Informa business

No claim to original U.S. Government works

Printed on acid-free paper

International Standard Book Number-13: 978-1-138-60681-4 (Hardback)

This book contains information obtained from authentic and highly regarded sources. Reasonable efforts have been made to publish reliable data and information, but the author and publisher cannot assume responsibility for the validity of all materials or the consequences of their use. The authors and publishers have attempted to trace the copyright holders of all material reproduced in this publication and apologize to copyright holders if permission to publish in this form has not been obtained. If any copyright material has not been acknowledged, please write and let us know so we may rectify in any future reprint.

Except as permitted under U.S. Copyright Law, no part of this book may be reprinted, reproduced, transmitted, or utilized in any form by any electronic, mechanical, or other means, now known or hereafter invented, including photocopying, microfilming, and recording, or in any information storage or retrieval system, without written permission from the publishers.

For permission to photocopy or use material electronically from this work, please access www. copyright.com (http://www.copyright.com/) or contact the Copyright Clearance Center, Inc. (CCC), 222 Rosewood Drive, Danvers, MA 01923, 978-750-8400. CCC is a not-for-profit organization that provides licences and registration for a variety of users. For organizations that have been granted a photocopy licence by the CCC, a separate system of payment has been arranged.

Trademark Notice: Product or corporate names may be trademarks or registered trademarks, and are used only for identification and explanation without intent to infringe.

Library of Congress Cataloging-in-Publication Data

Names: Parnell, Katie J., author. | Stanton, Neville A. (Neville Anthony), 1960- author. | Plant, Katherine L., author.
Title: Driver distraction : a sociotechnical systems approach / authored by Katie J. Parnell, Neville A. Stanton, and Katherine L. Plant.
Description: Boca Raton : Taylor & Francis, a CRC title, part of the Taylor & Francis imprint, a member of the Taylor & Francis Group, the academic division of T&F Informa, plc, 2019. | Series: Transportation human factors: aerospace, aviation, maritime, rail, and road series
Identifiers: LCCN 2018036626 | ISBN 9781138606814 (hardback : acid-free paper) | ISBN 9780429466809 (e-book)
Subjects: LCSH: Automobile drivers—Psychology. | Distracted driving—Prevention. | Automobiles—Design and construction. | Human-machine systems.
Classification: LCC TL152.3 .P36 2018 | DDC 363.12/57—dc23
LC record available at https://lccn.loc.gov/2018036626

Visit the Taylor & Francis Web site at
http://www.taylorandfrancis.com

and the CRC Press Web site at
http://www.crcpress.com

Katie J. Parnell

For Mum and Dad

Neville A. Stanton

For Maggie, Josh and Jem

Katherine L. Plant

For The HFE Dream Team

Contents

Preface..xiii
Acknowledgements xv
List of Abbreviations..xvii
Authors...xix

Chapter 1   Introduction 1

   1.1   Background...1
   1.2   Aim and Objectives 3
   1.3   Structure of the Book 4

Chapter 2   Driver Distraction, Technology and the Sociotechnical Systems
            Approach 7

   2.1   Introduction 7
   2.2   Driver Distraction..7
         2.2.1   In-Vehicle Technology Developments and Driver
                 Distraction 9
                 2.2.1.1   Crash Statistics..12
                 2.2.1.2   Legislation 14
                 2.2.1.3   Design Guidelines....................................15
   2.3   The Sociotechnical Systems Approach to Accident Analysis ... 16
         2.3.1   The Sociotechnical System Approach to Driver
                 Distraction 17
   2.4   Conclusion 18

Chapter 3   Driver Distraction Methodology 19

   3.1   Introduction 19
         3.1.1   Driver Distraction Methodological Challenges..........19
   3.2   Classification of Methodologies 21
         3.2.1   Objective Quantitative Methods: Measuring
                 Behaviours..30
         3.2.2   Objective Qualitative Methods: Observing
                 Behaviours..31
         3.2.3   Subjective Quantitative Methods: Measuring
                 Opinions 31
         3.2.4   Subjective Qualitative Methods: Observing
                 Opinions 32
   3.3   Systems Methodology and Driver Distraction 33
   3.4   The Way Forward 38
   3.5   Conclusion 39

vii

viii   Contents

Chapter 4 Exploring the Mechanisms of Driver Distraction:
The Development of the PARRC Model 41

  4.1   Introduction 41
        4.1.1   A Case of Driver Distraction 41
        4.1.2   What Causes Driver Distraction? 43
        4.1.3   Modelling the Sociotechnical System 45
  4.2   Methodological Approach 46
        4.2.1   Grounded Theory 46
                4.2.1.1   Document Analysis 47
  4.3   Results and Discussion 48
        4.3.1   Causal Factors 49
                4.3.1.1   Factor 1: Adapt to Demands 49
                4.3.1.2   Factor 2: Behavioural Regulation 50
                4.3.1.3   Factor 3: Goal Conflict 50
                4.3.1.4   Factor 4: Goal Prioritisation 51
                4.3.1.5   Factor 5: Resource Constraints 51
        4.3.2   Interconnections 51
  4.4   Application of the PARRC Model: A Case Study of
        Distracted Driving 54
        4.4.1   The Systems Approach and the PARRC Model 56
                4.4.1.1   Goal Conflict 57
                4.4.1.2   Resource Constraints 58
                4.4.1.3   Adapt to Demands 59
                4.4.1.4   Goal Priority 59
                4.4.1.5   Behavioural Regulation 60
  4.5   General Discussion 60
        4.5.1   Theoretical Implications 60
        4.5.2   Practical Implications 61
  4.6   Conclusion 62

Chapter 5 What's the Law Got to Do with It? Legislation Regarding
In-Vehicle Technology Use and Its Impact on Driver Distraction 63

  5.1   Introduction 63
        5.1.1   The Role of Legislation in the Road Transport
                Domain 63
  5.2   Method 65
        5.2.1   Application of the Risk Management Framework
                to In-Vehicle Technology Use 65
        5.2.2   Application of the AcciMap Analysis to
                In-Vehicle Technology Use 68
        5.2.3   AcciMap Analysis 69
  5.3   Results and Discussion 70
        5.3.1   AcciMap of Phone Use 70
        5.3.2   AcciMap of Other Technology Use 75

Contents   ix

5.3.3   Comparison between Mobile Phone Use
AcciMap and Other Technology Use AcciMap 77
5.3.4   Application to Specific Events 79
    5.3.4.1   Scenario 1 79
    5.3.4.2   Scenario 2 80
5.4   General Discussion 82
  5.4.1   Recommendations 84
  5.4.2   Evaluation and Future Research 84
5.5   Conclusion 87

Chapter 6   Creating the Conditions for Driver Distraction: A Thematic
Framework of Sociotechnical Factors 89

6.1   Introduction 89
  6.1.1   Voluntary Distraction: Theory and Methodology 89
  6.1.2   Objectives 92
6.2   Study 1 92
  6.2.1   Aim 92
  6.2.2   Method 92
    6.2.2.1   Participants 92
    6.2.2.2   Data Collection 93
  6.2.3   Data Analysis 95
    6.2.3.1   Inductive Thematic Coding 96
    6.2.3.2   Inter-Rater and Intra-Rater Reliability
Assessment 97
  6.2.4   Results 98
  6.2.5   Discussion 101
6.3   Study 2 103
  6.3.1   Aim 103
  6.3.2   Method 103
  6.3.3   Results 104
    6.3.3.1   Adapt to Demands 106
    6.3.3.2   Behavioural Regulation 106
    6.3.3.3   Goal Conflict 106
    6.3.3.4   Goal Priority 107
    6.3.3.5   Resource Constraints 107
    6.3.3.6   Interconnections 108
  6.3.4   Discussion 113
6.4   General Discussion 114
  6.4.1   Recommendations to Practise 115
    6.4.1.1   Driver 117
    6.4.1.2   Infrastructure 117
    6.4.1.3   Task 117
    6.4.1.4   Context 118
  6.4.2   Evaluation and Future Work 119
6.5   Conclusion 120

x   Contents

Chapter 7 What Technologies Do People Use When Driving and Why?......... 121

7.1   Introduction 121
      7.1.1   Factors Linked to Technology Engagement............. 121
7.2   Method... 123
      7.2.1   Semi-Structured Interview Study........................... 123
            7.2.1.1   Interview Participants............................... 123
            7.2.1.2   Interview Procedure 123
      7.2.2   Online Survey Study 124
            7.2.2.1   Online Survey Participants....................... 124
            7.2.2.2   Online Survey Procedure 125
      7.2.3   Data Analysis 125
7.3   Results and Discussion 126
      7.3.1   Interview and Online Survey Sample Correlation....126
      7.3.2   Likelihood Ratings.. 127
            7.3.2.1   Younger Age Category 128
            7.3.2.2   Middle Age Category 129
            7.3.2.3   Older Age Category 130
      7.3.3   Likelihood Reasoning 131
            7.3.3.1   Satnav... 131
            7.3.3.2   Hands-Free Phone.................................... 132
            7.3.3.3   In-Vehicle Infotainment Systems............. 134
            7.3.3.4   Voice Command System........................... 135
            7.3.3.5   Mobile Phone.. 137
            7.3.3.6   Road Type 139
            7.3.3.7   Legislation 140
7.4   Implications 140
7.5   Evaluation and Future Work... 144
7.6   Conclusion 145

Chapter 8 Good Intentions? Willingness to Engage with Technology on
the Road and in a Driving Simulator 147

8.1   Introduction 147
      8.1.1   Naturalistic Decision Making 147
      8.1.2   Experimental Setting... 149
8.2   Method... 151
      8.2.1   Participants... 151
      8.2.2   Experimental Design.. 152
      8.2.3   Equipment 152
            8.2.3.1   Vehicles.. 152
      8.2.4   Procedure 153
            8.2.4.1   Verbal Protocol Methodology.................. 153
            8.2.4.2   Task Scenarios 154
            8.2.4.3   Route... 156
            8.2.4.4   NASA-TLX.. 156
      8.2.5   Data Analysis 157

Contents   xi

8.3   Results 158
      8.3.1   Scenario Responses 159
            8.3.1.1   Pre-Trial Interview................................ 159
            8.3.1.2   Stated Intention..................................... 159
            8.3.1.3   Road Type and Task Type....................... 160
            8.3.1.4   Reasons for Stated Intention.................... 161
      8.3.2   Matrix Queries 163
      8.3.3   Driving Speed... 168
            8.3.3.1   Mean Speed 168
            8.3.3.2   Speed Variability 170
      8.3.4   Workload 170
8.4   Discussion.. 171
      8.4.1   Experimental Condition 171
      8.4.2   Factors Affecting Naturalistic Decision Making..... 171
            8.4.2.1   Task Type.. 172
            8.4.2.2   Road Type... 173
      8.4.3   Using Verbal Protocol to Capture Naturalistic
               Decision Making 174
      8.4.4   Future Research... 175
8.5   Conclusion 176

Chapter 9   Evolution of the PARRC Model of Driver Distraction:
Development, Application and Validation 177

9.1   Introduction 177
9.2   Developing the PARRC Model.................................. 180
      9.2.1   Stage 1. Grounded Theory: Model Development..... 180
      9.2.2   Stage 2. AcciMap Analysis: Model Application 181
      9.2.3   Stage 3. Semi-Structured Interviews: Model
               Validation 181
      9.2.4   Stage 4. Driving Study: Model Validation 182
9.3   A Sociotechnical Systems Definition of Driver Distraction... 185
9.4   Discussion.. 185
9.5   Recommendations 186
      9.5.1   International Committees, National Committees
               and Government 187
      9.5.2   Regulators... 188
      9.5.3   Industrialists 189
      9.5.4   Resource Providers... 190
      9.5.5   End Users 191
      9.5.6   Equipment and Environment................................... 192
9.6   Conclusion 193

Chapter 10   Conclusions 195

10.1   Introduction 195

| 10.2 | Summary of Findings | 195 |

10.2 Summary of Findings 195
10.3 Future Work 199
    10.3.1 Theoretical Implications 200
        10.3.1.1 Sociotechnical Systems Theory and Definition of Driver Distraction 200
        10.3.1.2 Further Exploration of the PARRC Model 200
        10.3.1.3 Cultural Differences 200
        10.3.1.4 Context and Road Environment 201
        10.3.1.5 Automated Driving 201
        10.3.1.6 Applications to Other Domains 203
    10.3.2 Methodological Implications 203
        10.3.2.1 Qualitative Research 203
        10.3.2.2 Experimental Setting 204
        10.3.2.3 Eye Tracking 204
    10.3.3 Practical Implications 205
        10.3.3.1 Recommendations to Practise for Alternative Countermeasures 205
10.4 Closing Remarks 205

Appendix A 207

Appendix B 217

References 221

Author Index 243

Subject Index 247

Preface

The work presented in this book explores the issue of driver distraction that is increasingly sourced from the array of technological devices that are now commonly found within the vehicle. The impact of driver distraction on road safety has been a focus of Human Factors research for more than half a century. Over this time, there have been many developments in the design of vehicles and methods applied to the study of driver safety. The information technology revolution is also changing the way drivers interact with devices in the vehicle. Whilst these technologies may enhance the drivers' experience, and even provide options to monitor and improve driving performance, there are concerns regarding safety. As the sources of distraction change with the integration of technology within the vehicle, as well as everyday life, the traditional approaches to managing driver safety need to be reviewed.

Contemporary methods for studying driver distraction have focused largely on the individual driver, identifying them as the source of error. This approach has aimed to reduce distraction by monitoring and penalising their behaviour through legislation and surveillance. Yet, within the accident analysis domain more generally, the sociotechnical systems approach is in ascendancy. This perspective focuses on analysis of the whole system, its values, and the interactions between human and technical elements at all organisational levels. This approach provides holistic and novel countermeasures to prevent adverse events from occurring in the future. This book highlights the importance of the systems approach to the domain of driver distraction, which until now has been heavily focused on the individual driver. As technology advancements are likely to continue unabated, and driver distraction is one of the top five dangerous road-related behaviours, now is the time to adopt a systems perspective.

This book provides theoretical, methodological and practical contributions to the field of driver distraction. It is aimed at practitioners involved in accident and incident investigation as well as researchers in the field of Road Safety, Human Factors, Ergonomics, Engineering and Psychology. For road safety practitioners, we recommend the methods and data presented in this book as well as the practical recommendations. For researchers and students, we recommend the theoretical concepts and studies presented here to stimulate further research in the area.

Acknowledgements

We would like to acknowledge the funding received from the Engineering and Physical Science Research Council and the sponsorship received from Jaguar Land Rover in facilitating the work presented in this book. We would also like to thank all participants who kindly volunteered their time and contributed to the data collected within this research. Thanks also goes to Dr Alexander Eriksson for his assistance in conducting the online trials as well as to Dr Craig Allison and James Brown for their assistance in developing the driving simulations. We are also very grateful for the reviewers for taking the time to review the book and for their positive feedback. Finally, many thanks to Poppy Robinson for creating the design for the cover of the book.

List of Abbreviations

ACC	Adaptive Cruise Control
CR	Correct Rejection
DfT	Department for Transport
FA	False Alarm
HFACS	Human Factors Analysis and Classification System
HMI	Human-Machine Interface
Hr	Hour
ISO	International Organisation for Standardisation
IVIS	In-Vehicle Infotainment System
MRT	Multiple Resource Theory
Mph	Miles per Hour
NDM	Naturalistic Decision Making
NHTSA	National Highway Traffic Safety Administration
PARRC	Prioritise, Adapt, Resource, Regulate, Conflict
PDT	Peripheral Detection Task
RMF	Risk Management Framework
SAE	Society of Automotive Engineers
SD	Standard Deviation
SHRP	Strategic Highway Research Program
STAMP	Systems Theoretic Accident Model and Process
TCI	Task Capability Interface
TRB	Transportation Research Board
VP	Verbal Protocol
Yr	Year

Authors

Dr Katie J. Parnell, BSc, EngD, is a post-doc research fellow at the University of Southampton. She studied for her Engineering Doctorate at the University of Southampton which focused on the topic of driver distraction from in-vehicle technology within automotive vehicles. She also earned a first-class BSc Psychology degree from the University of Reading. Her research interests include applying, developing and reviewing accident causation from a sociotechnical systems viewpoint. Katie is also interested in the advancements in technological interfaces and how they can be utilised safely. She has published a number of journal articles on applying the sociotechnical systems approach to the study and mitigation of driver distraction. In her current role as a post-doc research fellow at the University of Southampton, she is now studying the integration of novel technologies into the cockpits of aircraft within the Open Flight Deck project.

Professor Neville A. Stanton, PhD, is a Chartered Psychologist, Chartered Engineer and a Chartered Ergonomist and holds the Chair in Human Factors in the Faculty of Engineering and the Environment at the University of Southampton. Professor Stanton is the founder and Director of the Human Factors Engineering team in the Transportation Research Group. He has degrees in Psychology, Applied Psychology and Human Factors and has worked at the Universities of Aston, Brunel, Cornell and MIT. Professor Stanton has supervised 27 doctoral students to completion, four of which now hold the title of Professor. His research interests include modelling, predicting and analysing human performance in transport systems, as well as designing the interfaces between humans and technology. Professor Stanton has worked on cockpit design in automobiles and aircraft over the past 25 years, working on a variety of automation projects. He has published 40 books and over 300 journal papers on Ergonomics and Human Factors and is currently an editor of the peer-reviewed journal *Ergonomics*. In 1998, he was awarded the Institution of Electrical Engineers Divisional Premium Award for a co-authored paper on Engineering Psychology and System Safety. The Chartered Institute of Ergonomics and Human Factors awarded him the Otto Edholm Medal in 2001, The President's Medal in 2008 and 2018, and the Sir Frederic Bartlett Medal in 2012 for his contribution to basic and applied ergonomics research. The Royal Aeronautical Society awarded him and his colleagues the Hodgson Prize and Bronze Medal in 2006 for research on design-induced flight-deck error published in *The Aeronautical Journal*. The University of Southampton has awarded him a Doctor of Science in 2014 for his sustained contribution to the development and validation of Human Factors methods.

Dr Katherine L. Plant, BSc, PhD, is a Lecturer in Human Factors Engineering in the Transportation Research Group within the Faculty of Engineering and the Environment at the University of Southampton. She is the technical lead for aviation and road safety research within the group. In 2014, Dr Plant was awarded the Honourable Company of Air Pilots Prize for Aviation Safety for her

xix

research exploring aeronautical critical decision making. Her primary research interests centre on understanding how the interaction of the environment that we work in and the mental schema that we hold influence our actions and decision-making processes. Dr Plant is passionate about teaching Human Factors and runs the module 'Human Factors in Engineering', which is offered to under- and postgraduate engineering students across the faculty. In addition to this, Dr Plant supervises a number of PhD, postgraduate and undergraduate student projects.

1 Introduction

1.1 BACKGROUND

Since the invention of the modern motor car in the late 1800s and its widespread use from the early 1900s, there have been many developments to the road environment, infrastructure, vehicles, in-vehicle technology, licensure and driver training. The advancement of the road transport system through the years has greatly benefited society, yet road safety has become a critical area of concern as it is threatened by a number of issues, predominately the fatal five: drink-driving, not wearing a seat-belt, fragility of motorcycle helmets, speeding and driver distraction (WHO, 2016). Driver distraction, in particular, is a very complex area that has been the subject of academic research and public concern for over 50 years (e.g. Brown, 1965), yet no universal definition or approach to the issue has been achieved (e.g. Young et al., 2008a; Regan et al., 2011). This is due to some contention within the research field on numerous aspects of the behaviour, including its distinction from driver inattention (Regan et al., 2011), its intentionality (Young et al., 2008) and its dependence on its context of occurrence (Lee, 2014). It is suggested that driver distraction is a form of driver inattention that involves the diversion of attention away from safety critical activities within the driving task (Regan et al., 2011; Young et al., 2008). Yet, the mechanisms through which attention is diverted (Young & Salmon, 2012) and the activities that are determined to be 'safety critical' are still unclear (Trick et al., 2004; Regan et al., 2011).

Many have attempted to provide a definition of driver distraction (e.g. Treat, 1980; Manser et al., 2004; Sheridan, 2004; Pettitt et al., 2005; Young et al., 2008; Hoel et al., 2010; Regan et al., 2011; Hedlund et al., 2005). The multiple attempts to define the behaviour have, however, led to diverging avenues of research that make comparisons across the literature difficult and require caution when assessing estimates and outputs of distraction (Gordon, 2008; Regan et al., 2008, 2011). The multiple sources of distraction, be those that are voluntary or involuntary, inside or outside the vehicle, or driven by events versus internal thoughts, complicate the study and definition of the behaviour.

Technological advancement has played an increasingly important role in the progression of the automotive domain. Burnett (2009) states how technologies have contributed to the development of vehicles through the provision of information-based systems, control-based systems and other functionalities. Some information- and control-based systems support the driving task by providing driving-relevant information, for example satnav systems, and automated tasks that assist the driving task, for example cruise control. Yet, other functionalities that do not directly assist the driving task, and may actually adversely affect the drivers' safe monitoring of the driving task, have also been the product of rapid technological advancement, for example music players and entertainment systems (Walker et al., 2001; Stutts et al., 2001; Burnett, 2009; WHO, 2011). In-vehicle systems now provide drivers with an array of information, entertainment, communication and comfort features to enhance the driving experience (Burnett, 2009; Harvey & Stanton, 2013). As technology has developed, so have the variety and complexity of these features (Walker et al., 2001) and the prevalence of nomadic devices that drivers bring in to the vehicle (Janitzek et al., 2010). While the statistical data related to crash risk is difficult to discern (Sullman, 2012), research has shown that music devices (Lee et al., 2012), satnav systems (Tsimhoni et al., 2004), wearable technologies (e.g. Sawyer et al., 2014) and hands-free devices (Horrey & Wickens, 2002) impair drivers' attention in the driving task.

Legislation and regulations must adapt to incorporate technological distractions, yet there is critique that policy change may be somewhat of an afterthought, playing catch-up only after gaps within the existing policy have been found (Leveson, 2011). With developments in technology occurring at a rapid pace, it is hard for policy to regulate its use (Leveson, 2011; Redelmeier & Tibshirani, 1997). This is evidenced by the recent (March 2017) change in UK legislation that raised the number of penalty points from 3 to 6 (where 12 points incurs a ban from driving for a year) and increased the fine to £200 if drivers are caught using their handheld mobile phone while driving. This is in response to the continuation of the issue surrounding phone use while driving. Previous penalty increases were made in 2013 to £100 from the £60 fine in 2007, which was in turn increased from the initial penalty of £30 set in 2003. The continual need to increase these penalties suggests these measures are ineffective in their current form.

There is a claim within the road safety literature that techniques which focus on penalising the driver descend from a traditional or 'old view' (Reason, 2000; Dekker, 2002) of accident causation, that view the driver as unreliable and the main threat to safety (Larsson et al., 2010). This is opposed to the 'new' systems approach that considers accident causation to be a consequence of the interrelationships within the sociotechnical system (e.g. Larsson et al., 2010; Leveson, 2011; Salmon, McClure & Stanton, 2012; Lansdown et al., 2015). Von Bertalanffy (1968) proposed that complex systems value the interactions between multiple elements that comprise a system, as well as the wider environment within which they are located, rather than focusing on individual elements. The systems approach to road safety behaviour provides an interesting view as it assesses the behaviour of the driver with respect to the wider context within which it occurs, as well as the identification of other actors within the environment that influence it. The potential for the application of a systems perspective to driver distraction has been suggested in recent years (Tingvall et al., 2009; Young & Salmon, 2012, 2015). Yet, the rapid development of technologies that have been placed and integrated into the vehicle is thought to increase the need to review the behaviour from this perspective (Dekker, 2002). The application of the sociotechnical perspective to the study of driver distraction from technological devices is therefore the focus of this book. The relevance of the approach to the domain will be considered, as well as its ability to understand the problem, identify key issues and provide recommendations for novel solutions. It should be noted that the focus of this research is on passenger cars, rather than commercial and freight drivers who have their own associated problems with fatigue and inattention that

Introduction 3

stem from their extended hours on the road and associated sleep deprivation (Bunn et al., 2005; Hickman & Hanowski, 2012).

1.2 AIM AND OBJECTIVES

The main aim of this book is to explore the possibilities that taking a sociotechnical systems approach to driver distraction from technological devices can have on understanding the behaviour and how to mitigate it. To achieve this aim, three main objectives have motivated the work that has contributed to the development of this book. These include the following:

1. Review the current methodologies that are used to study driver distraction and seek out methods that can assist in the application of the sociotechnical systems theory to the study of driver distraction.

   The study of driver distraction has been around for a number of decades. Across this time period, a host of different methodologies have been applied to measure, understand and provide possible countermeasures to the behaviour. Yet, as there has been little focus on the sociotechnical system within which driver distraction occurs, the application of methods that can account for the wider system in the underlying cause and consequences of driver distraction will be discussed and applied. This aims to move away from the traditional individual-centric methods that highlight the importance of individual actors in the cause of distraction towards a more holistic approach with the use of methods that facilitate a wider systems-based approach. This book reviews the previous methods that have been used to study driver distraction and looks to apply novel systems-based methods that can provide unique insights into the behaviour.

2. Develop a framework through which to study driver distraction from a sociotechnical systems approach.

   In order to explore the possibilities that taking a sociotechnical systems approach to the study of driver distraction can have, a framework was sought that can account for the key factors of distraction and the relationship that they have to the sociotechnical system in which it occurs. The development of such a framework will assist in structuring the study of the behaviour and enable it to inform future recommendations that incorporate systems thinking. While there have been many attempts to model driver distraction, no universal approach has been attained. The addition of a sociotechnical systems approach to the literature aims to increase the scope for a more holistic view of driver distraction going forward.

3. Assess the current countermeasures that are used to tackle driver distraction from technological devices and determine the potential for novel countermeasures.

   The background to driver distraction that was given at the beginning of this chapter highlighted that the current focus on increasing the penalties that drivers receive when they are caught engaging in distracting tasks while driving may not be effective in negating the issue on their own.

The effectiveness of legislation in tackling driver distraction will be explored as well as exploring possible alternative countermeasures to driver distraction, which may provide more effective avenues for distraction mitigation strategies. This will require an assessment of how current legislation is currently operating, an understanding of why drivers choose to engage with tasks that may pose as potential distractions and possible ways in which this can be mitigated to maintain road safety.

1.3 STRUCTURE OF THE BOOK

This book is written to inform the reader of the current state of knowledge on driver distraction and the important implications of taking a sociotechnical systems approach to the problem. In doing so, it discusses the development of a model of driver distraction by the authors that is validated using different research methodologies across the chapters of the book. To this end, the chapters of the book follow on from one another, with each building on the work of the previous. Yet, the chapters are also written so that they can stand alone, to be read individually where this is required. A brief overview of the contents of each chapter is given below.

Chapter 1: Introduction

This first chapter introduces the background to the research area and the work conducted in the production of this book. It will present the overall structure of the book and its contribution to the theory, methodology and practice surrounding driver distraction.

Chapter 2: Driver Distraction, Technology and the Sociotechnical Systems Approach

The second chapter discusses the previous applications of Human Factors methods and approaches to accident analysis to understand how this has informed knowledge on the distractive effect of technological development. A brief history of the developments in in-vehicle technology is detailed. The current challenges in determining the impact of the issue using crash statistics is then discussed before reviewing the current ways of mitigating against distraction, including legislation and design guidelines. This chapter then provides a detailed introduction to sociotechnical systems theory and the use it may have to the study of driver distraction.

Chapter 3: Driver Distraction Methodology

This chapter discusses the methods that have been used to study driver distraction and the resulting conclusions that they have enabled. A research-practice gap is suggested between the methods that have been applied and the practices that are in place to mitigate the behaviour. The movement towards the sociotechnical view of safety across other domains and the associated methodologies that have been used to reveal novel insights and solutions to safety maintenance are discussed in relation to driver distraction.

Chapter 4: Exploring the Mechanisms of Distraction from In-Vehicle Technology: The Development of the PARRC Model

This chapter builds on the identified need to apply the sociotechnical perspective to driver distraction from in-vehicle technology from Chapter 2 and sets out to observe the factors involved in technological distractions. It applies a grounded

Introduction

theory methodology to explore the factors that have been identified in the literature and incorporates these into the Prioritise, Adapt, Resource, Regulate, Conflict (PARRC) model of driver distraction from in-vehicle technology. The application of this model to a case study of real-world distraction demonstrates the ability of the model to account for the responsibility of systemic actors through the identified PARRC model factors in the lead up to the incident.

Chapter 5: What's the Law Got to Do with It? Legislation Regarding In-Vehicle Technology Use and Its Impact on Driver Distraction

This chapter explores the legislation that is currently in place in the UK that targets driver distraction from in-vehicle technology. This seeks to identify all of the actors that the legislation implicates, the decisions that these actors make in response to the legislation and the influence that this has on the mitigation of driver distraction. A distinction in legislation is identified between that specifically targeting handheld mobile phone use and that incorporating all other technologies. The implications that are revealed from taking a sociotechnical systems view are identified within an AcciMap analysis that contrasts the actors and actions that relate to the distinction that is set out within legislation. This revealed the range of systemic actors that are implicated by the legislation and the proposition that current laws may be focusing heavily on preventing distraction from handheld mobile phones while creating the conditions for distraction from other technologies. Application of the PARRC model, developed in Chapter 4, to the AcciMap analysis suggests how previous limitations of the methodology may be overcome, as well as identifying the mechanisms through which legislation may be guiding driver distraction.

Chapter 6: Creating the Conditions for Driver Distraction: A Thematic Framework of Sociotechnical Factors

This chapter presents the method used to conduct and analyse a semi-structured interview study with 30 drivers to establish the reasons why they may be more or less likely to engage with a variety of different technologies while driving across different road types. The qualitative data was analysed to obtain the factors that drivers state influence their decision to engage with technology while driving. The chapter is split into two studies. The first presents the inductive thematic analysis that is conducted on the open-ended discussions which enabled the development of an extensive thematic framework of the factors that drivers state influences their use of technology while driving. The second study details the procedure of applying the factors generated by the drivers to the factors identified from the literature, and how they relate to the PARRC model that was developed in Chapter 3. The findings of both studies are discussed in relation to the current literature, and implications for future recommendations in targeting driver distraction are given.

Chapter 7: What Technologies Do People Use When Driving and Why?

This chapter presents further analysis of the data generated from the semi-structured interviews detailed in Chapter 6. It will expand on how, why and when drivers choose to engage with a range of technologies and the influence of age, gender and road type on these decisions. This will include the analysis of the quantitative data generated from asking drivers to rate their likelihood of engaging with different technological tasks across road types. The correlation of these quantitative ratings to a larger online survey is used to assess the generalisability of the findings.

Qualitative analysis is also used to explore the reasons drivers gave for their likelihood ratings. These were coded to the thematic framework developed in Chapter 6 and are explored across age and gender categories.

Chapter 8: Good Intentions: Willingness to Engage with Technology on the Road and in a Driving Simulator

This chapter presents an experimental study that was conducted to assess the decisions that drivers make to engage with technologies while they were driving within a real-world environment and a driving simulator environment. It builds on the semi-structured interview study discussed in Chapters 6 and 7 by asking drivers to state their intention to engage with technology in predefined scenarios and the decision-making process that led to their intention, by providing verbal protocols whilst driving. This aims to understand when drivers may engage with technological tasks, how they may go about doing it and why. The different experimental environments, on-road and simulator, are compared within the data analysis. The impact that the drivers' provision of verbal protocols has on speed metrics in these environments is also assessed.

Chapter 9: Evolution of the PARRC Model: Development, Application and Validation

The penultimate chapter provides an overview of the PARRC model of distraction, from its conception in Chapter 4, using grounded theory methodology, to its application to the thematic framework developed from the semi-structured interview data in Chapter 6, before introducing the insights that were gained from the simulator and road experiments presented in Chapter 8. Findings and recommendations that were identified across the evolution of the PARRC model throughout the book in relation to the sociotechnical system surrounding the behaviour are discussed. The insights that were gained from the study of driver distraction from technological devices across the work presented within the book were also used to generate a sociotechnical systems definition of the behaviour.

Chapter 10: Conclusions

This chapter concludes the book by summarising the findings that have been generated through the research conducted in the production of this book. These are discussed in relation to the objectives of the book outlined in this chapter. The key conclusions relate to the insights that can be obtained through the application of the sociotechnical systems approach. A research-practise gap is identified within accident analysis in the road transport system which suggests that while there is a shift in research towards adopting a more systems-based approach, this is not represented in the practises used to maintain safety, especially in relation to driver distraction. Therefore, the methods applied in this book and the systems-based framework that is developed aim to demonstrate how research into the sociotechnical systems approach can translate into practise by providing systems-based recommendations that target all elements responsible for driver distraction. The implications of the work including its theoretical, methodological and practical applications are provided.

2 Driver Distraction, Technology and the Sociotechnical Systems Approach

2.1 INTRODUCTION

Human Factors, as a discipline, largely focuses on the application of methodologies to solve practical problems (Stanton et al., 2013). Human Factors methods have traditionally been applied to four main areas: design, training, safety assessment and accident investigation (Cacciabue, 2013). This has enabled the application of Human Factors to many safety critical domains, such as aviation (Plant & Stanton, 2016), healthcare (Reason, 1995), maritime (Baber et al., 2013), rail (Stanton & Walker, 2011) and road transport (Stanton & Salmon, 2009). Human Factors has long been concerned with the recognition, assessment and analysis of error and its relationship to incidents (Reason, 1990). Within the road transport domain, there is a suggestion that the view of error causation and management may be lagging behind other domains, such as aviation (Stanton & Salmon, 2009; Salmon et al., 2010). This is attributed to a lack of appropriate methods to measure and classify the errors that are made within the road transport domain which, in turn, results in a lack of understanding for the causal factors that lead to error (Stanton & Salmon, 2009).

This chapter will introduce the current state of understanding on driver distraction including the current issues in defining the behaviour and its multiple sources. The technological sources of distraction are then given in more detail, as this is the main focus of the book. The rapid advancement of in-vehicle technology has had large-scale implications for the issue of driver distraction (Harvey & Stanton, 2013). These are discussed with regard to the legislation that is currently in place to manage the issue. The crash statistics that strive to determine the magnitude of the issue and the design guidelines that have been established to regulate the distractive effects of technologies within the vehicle. Finally, the sociotechnical systems approach is expanded on and the benefits that it can have to the study of driver distraction from technological sources of distraction in the vehicle are discussed.

2.2 DRIVER DISTRACTION

Driver distraction is an activity that features within many efforts in the literature to classify driving errors (e.g. Reason et al., 1990; Wierwille et al., 2002; Sabey & Taylor, 1980). A review of human error taxonomies in the road transport domain have, however, suggested that distraction is not an error in itself but that errors occur as a result of being distracted (Stanton & Salmon, 2009). An issue with the current approach to driver distraction is that it is often only with hindsight that distraction is recognised, once adverse events have occurred, only then does it become evident what the driver was, or was not, paying attention to and how the incident could have been avoided (Kircher & Ahlstrom, 2016). This suggests that on many occasions drivers can divert their attention away from the road without this having an impact on the driving task or resulting in incident (Kircher & Ahlstrom, 2016).

This is one of the issues that makes defining driver distraction problematic. Other issues include its relationship to driver inattention (Regan et al., 2011), with some considering distraction to be a form of inattention (e.g. Pettit et al., 2005; Victor et al., 2008) or a taxonomic sublevel of inattention (Regan et al., 2011), while others classify it as a separate entity to distraction (e.g. Treat et al., 1980). These views differ on subtleties that reflect the mechanisms through which distraction is deemed to occur. Cnossen et al. (2000) highlight the important differences between approaches that focus on inattention as the cause of distraction and those that focus on the mechanisms drivers use to incorporate the primary and secondary driving tasks together. Inattention suggests secondary tasks will invariably result in a failure to attend to the driving task, having undesirable effects on performance. Others suggest a strategic application of attention while driving with a secondary task to manage increased demand (Cnossen et al., 2000; Kircher & Ahlstrom, 2016).

The multiple definitions of driver distraction have been highlighted in the literature (Young et al., 2007; Regan et al., 2011), and while there are differences between them, there are common points that are included across the attempts that have been made. These include the notion of a 'diversion of attention', which is also the Oxford English definition of the term 'distraction' (Shorter Oxford English Dictionary on Historical Principles, cited in Regan et al., 2011). A competing activity that does not relate to, or that may impede on, the drivers' attention towards the driving task is also commonly referred to (e.g. Lee et al., 2008; Pettit et al., 2005; Hedlund et al., 2005; Ranney et al., 2000). Yet, there are discrepancies in the source, intentionality, process and outcomes that are referred to across definitions (Lee et al., 2008).

Distraction is typically categorised into four main subtypes: visual, cognitive, auditory and biomechanical (Young & Regan, 2007). Such categorisation identifies the different sources through which the drivers' attentional resources are diverted away from the driving task. The Multiple Resource Theory (MRT; Wickens, 2002) dictates that where two tasks compete for the same limited attentional resources, the interference between the two tasks affects their ability to be completed concurrently. The driving task is inherently a visual-spatial manual task that requires the driver to pay attention to the road scene as well as control the vehicle; therefore, the tasks that demand a high level of visual-spatial manual resources to complete are likely to impact on the safe completion of the driving task. Yet, highly demanding cognitive (Harbluk, Noy et al., 2007) and auditory tasks (Brodsky, 2002) are also a concern.

There are also differing opinions on the type of task that is classed as a distraction. Some argue that it is only those tasks which do not relate to driving that are distractions (e.g. Hedlund et al., 2005), whilst others suggest that a distraction may be any activity that diverts attention away from those essential to driving safely

Driver Distraction and Sociotechnical System

(Lee et al., 2008; Pettit et al., 2005; Drews & Strayer, 2008). Furthermore, the actor within the description of distraction is also a source of variability. The majority of definitions state that the driver is the source of the distraction, for example Patten et al. (2004) state that *"driver distraction implies that drivers do things that are not primarily relevant to the driving task"* (p. 342). Pettit et al. (2005) state that distraction is caused by a *"delay by the driver in the recognition of information necessary to safely maintain the lateral and longitudinal control of the vehicle"* (p. 11).

These definitions are indicative of an individual focused approach to driver distraction, which has been the predominant approach in the literature (Young & Salmon, 2015). Yet, some accounts have defined distraction as independent from the driver. For example, Manser et al. (cited in Young et al., 2008) stated that driver distraction is the result of *"objects or events both inside and outside of the vehicle that serve to redirect attention away from the task of driving or capture enough attention of the driver such that there are not enough attentional resources for the task of driving"* (p. 33), thus suggesting that distraction emerges from the competing activity rather than the driver. Drews and Strayer (2008) also suggest distraction does not stem from the driver diverting their attention but that it is the outcome of *"any event or activity that negatively affects the ability of the driver to process information that is necessary to safely operate a vehicle"* (p. 169). The sources of driver distraction and the involvement of the driver are central to this book. If we are to focus on the systems approach to driver distraction, then it is important that the definition of the behaviour does not focus responsibility solely on the driver but that the other actors within the wider system are also recognised. A systems definition of driving distraction has not yet been presented within the literature. This is something that the authors seek to rectify.

2.2.1 IN-VEHICLE TECHNOLOGY DEVELOPMENTS AND DRIVER DISTRACTION

We are said to be in an information technology revolution, which refers to the rapid developments in information and communication technologies that are providing innovative ways in converging computing and telecommunication to provide information across new and multifaceted platforms. In *Informational Technology and Organisation Change*, Eason (1988) states that the significance of technological advancement is the enhanced levels of information that it allows us to access, as information is the basis of human endeavour. Enhanced access to information gives us more options and enhances our decision-making process. Yet, we must be aware of the human and organisational changes that often accompany technological change. There are many advantages and benefits to the informational technology developments that have occurred, yet their ramifications within the wider society upon which the developments are having a direct impact on must be carefully considered.

The information technology revolution is apparent in the automobile vehicle industry; indeed, the automobile industry has co-evolved with modern developments in information technology (King & Lyytinen, 2005). Together they have progressed to provide entertainment-, communication- and performance-related information to the driver. The first technological advancement within the vehicle was the car radio, which was initially placed in the motor car in 1919 but became widespread in the industry in 1930 with the development of the company Motorola. This first initiated the idea that the vehicle could be a platform for providing the driver and passengers with entertainment while they drove. Since then developments in infotainment systems have aligned with advances in informational technology with cassette tape decks, CD players, stereo systems and MP3 players being integrated into the vehicle as standard. Communication devices have also become a feature within the vehicle. The first telephone was placed in the vehicle in the 1940s; in actual fact, cellular mobile phones were originally developed for use in the vehicle to be used by taxi drivers, provide traffic information and for other business purposes. Car phones were replaced by handheld phones which were invented in the 1970s but become popular into the 1980s and 1990s. Yet, as mentioned previously, the ramifications for technological advancement are little considered at the advent of individual systems. The development of mobile phones did not forecast the rise in their popularity and the impact that they would have on the development of society. The dangers of mobile phones in the vehicle was realised in the late 1990s, and laws began to come into place against their use by drivers in the 2000s. Other functionalities remain such as hands-free technology and integrated communication and entertainment devices built into the in-vehicle infotainment system. Nowadays, connected platforms link to smart phone devices to enable multiple applications and sources of information to be provided to the driver. There have also been developments in wearable technologies such as glasses and watches that provide additional platforms to communicate information to the user. As technologies have become increasingly popular and depended upon in modern society, the impact that they have on road safety must be realised.

It should be noted that not all developments and applications of technology in the vehicle pose a threat to road safety. As stated previously, developments in tasks that assist with the driving task, such as satellite navigation systems can provide improvements to traditional methods, such as paper maps (Srinivasan & Jovanis, 1997; Burnett & Joyner, 1997). Driver monitoring systems have also developed to try and determine when the driver is not attentive to the road and to bring their attention back to maintain their safe monitoring of the environment (Liang et al., 2007; Kutila et al., 2007; Dong et al., 2011 for a review). These systems offer potential for supporting the driver and helping them to remain safe (Dong et al., 2011). This is especially important with the developments in automation that may allow the driver to take more of their attentional resources away from the driving task but will require re-engagement when the road environment and automation require it (SAE, 2016; Stanton et al., 1997). System monitoring will enable warnings to determine when the driver needs to re-engage with the driving task, and ensure that they do so effectively (Gonçalves & Bengler, 2015). These systems are, however, challenged with enabling high levels of accuracy and being accepted by drivers (Dong et al., 2011). To avoid false alarms and allow drivers to trust and therefore use the system, hybrid methods of driver monitoring alongside assessment of performance metrics and information from technologies themselves are proposed as a useful combination in targeting driver distraction (Dong et al., 2011).

Some research in the field has suggested that the use of technologies while driving may actually have protective, rather than distractive, effects (Hickman & Hanowski, 2012; Caird et al., 2014). A naturalistic study into truck drivers found that when drivers were talking/listening on the phone, they actually had more time with eyes forward on the road and were more attentive to salient events in the road scene (Hickman & Hanowski, 2012). Interactions with the phone were not found to have the same effect (Hickman & Hanowski, 2012). This is similar to others who have suggested the possible benefits of passengers to the drivers' alertness and engagement in the driving task (Klauer et al., 2006; Olsen et al., 2009), although this is a complex and variable effect that is not always evidenced (Caird et al., 2014). The protective effect of talking on the phone or to a passenger is found more in naturalistic studies than in simulator research (Carsten & Merat, 2015). A review of the research concluded that while talking on a phone may minimise the number of less-severe rear-end crashes as drivers direct their eyes ahead, there is an increased risk of more serious accidents as drivers are less attentive to other road users and the wider road environment (Carsten & Merat, 2015). Furthermore, the research stemming from truck drivers cannot be related to passenger car vehicles as the extended periods of driving, levels of fatigue, training and journey types do not correspond to passenger vehicles, the latter of which is the focus of this book. Analysis of crash reports found mobile phones do not have protective effect (Redelmeier & Tibshirani, 1997).

There are, however, other benefits to having the technology available as needed in the vehicle if unanticipated events occur. For example, if a driver is involved in an accident or their vehicle breaks down, then access to a mobile phone is a necessity to allow them to get help when they are not driving. Therefore, banning mobile devices from entering the vehicle may not be advised, but preventing their misuse while driving the vehicle is an issue that needs to be tackled.

Listening to music in the vehicle is another task that has been linked to improving driver alertness and preventing fatigue (Cummings et al., 2001). In-vehicle music systems have advanced significantly over the past few decades, with progression from cassettes to CDs, to MP3 players and phone connectivity that allows smart phone applications to host an array of listening options for the driver (e.g. Apple Play). Other research has suggested that the type of music that drivers listen to influences their driving style and the safe control of the vehicle (Brodsky, 2002; Brodsky & Slor, 2013). This is another consideration when discussing the facilities that in-vehicle technologies provide and their influence on the safe control of the vehicle.

Despite these cases of technology development that suggest some benefits to preventing driver distraction, there are various other sources that identify the negative impact of technological device advancement on the safety of the driving task. Dekker (2011) highlights the issue of 'unruly technologies' in complex systems, whereby technological advancement is outgrowing the theories and countermeasures that can manage them. The 'unruly' aspect of technology refers to their unanticipated emergent behaviour through interaction with other elements in complex systems that are not adequately controlled for (Dekker, 2011). Technological developments are largely driven by industrial or commercial requirements which, Dorf (2001) claims, are harnessed by mankind to change or manipulate their environment. Advances in technology have facilitated a competitive relationship between manufacturers of wireless devices, computers and automobiles that has helped to propel the implementation of advanced devices such as Intelligent Transport Systems, e-mail servers and eco-driving systems into vehicles (Ranney et al., 2000). This satisfies the needs

12   Driver Distraction

of drivers who are becoming more reliant on the conveniences that wireless technologies offer (Dingus et al., 2006). Indeed, it is suggested that drivers are seeking to spend more of their productive time in the vehicle and enhance their enjoyment of the driving task (Burnett, 2009). Emphasis by manufacturers on reducing time-to-market has led system developers across industries to implement new technological developments at a rapid pace, without fully testing their effects on the system as a whole or how they may be effectively integrated (Leveson, 2011). Hence, developments may align with perceived consumer needs rather than their human capabilities.

It is therefore important to identify what the current issues are in relation to technology uptake in the vehicle and how they are impacting on distraction. The next sections will detail the current difficulties in the area that include capturing the size of the issue with the use of crash statistics. The use of legislation that specifically seeks to target technological distraction will then be discussed, alongside the design guidelines that aim to minimise the distraction through effective design of the technological interfaces.

2.2.1.1 Crash Statistics

Assessing the direct impact of distraction on road safety through accident statistics is challenging. There have been multiple attempts to establish the percentage of crashes that are due to driver distraction (e.g. Wang et al., 1996; Stutts et al., 2001; Neale et al., 2002; Stevens & Minton, 2001). Yet, due to there being no clear definition of what distraction is, diverging ways in which to measure and capture the behaviour have emerged which cause difficulties in reporting and recording distraction statistics. Retrospective analysis on the cause of accidents is difficult to determine. Legislation against the use of handheld phones has meant that data on their use by drivers can be collected through incident reports. Although this is still thought to underestimate the size of the issue due to a lack of witnesses as well as the drivers' own admission due to the legal implications (McCartt et al., 2006). Phone company billing information can be used to assess phone records at the time of the crash and then form as evidence for the use of phones while driving (McCartt et al., 2006). Yet, the use of other technologies in the vehicle and their propensity to cause incidents is not reported in the same way.

There is comparatively little evidence from crash statistics on the detrimental effects of the use of technologies other than mobile phones. This is attributable to a lack of data collected on the use of these devices such as wearable devices, infotainment systems and portable systems as they are not covered by such specific legislation. They are also hard to capture and infer causality. One study was able to capture some relevant data; Stevens and Minton (2001) were able to assess the accidents that arise due to entertainment systems in the vehicle. This revealed them to be the second largest cause of road accidents in the UK at the time of the report. Yet, another limitation of the use of crash data is its reliance on data from a set timeframe, and therefore the impact of novel technologies is also hard to determine (Stevens & Minton, 2001).

Beanland et al. (2013) used data from the Australian National In-depth Survey, which aimed to overcome some of the limitations of using crash reports by conducting in-depth interviews with those who had been reported to be involved in a crash

Driver Distraction and Sociotechnical System

to gain more detail about the circumstances surrounding the incident. This is useful when assessing the exact role that distraction played in the incident. Confidentiality of the interviews aimed to allow the individual to offer more information about the incident than they would in a police interview. The findings highlighted that internal distractions were the most prevalent, with visual-manual tasks particularly prominent. While in-vehicle systems and mobile phones featured as prevalent sources of distraction, other devices were not. Beanland et al. (2013) highlight that this is due to the time that the survey was conducted, in the year 2000 when nomadic devices were less prevalent. Thus, reporting methods and timescales mean that the use of crash data to determine the scale of distraction-related incidents from technology and their impact on road safety are not up to date and therefore unreliable. The 'unruly' nature of technologies (Dekker, 2011) and their rapid development mean that the use of crash statistics to provide up-to-date information on technology-based distractions is limited.

An alternative method to establish the prevalence of drivers' phone use is with naturalistic driving studies. These studies provide an opportunity to study drivers' behavioural patterns as well as the contextual factors that surround the behaviour. While they may be resource heavy and subject to the drivers being aware that they are being monitored, naturalistic driving studies have generated some important findings in relation to the driver and their use of technology while driving. The 100-car study was a large-scale naturalistic study that captured 109 drivers' behaviour over a 12- to 13-month period (Neale et al., 2002; Stutts et al., 2001; Klauer et al., 2006; Dingus et al., 2006). The large scale of the study aimed to capture crashes, near crashes and incident events to determine their contributing factors. This highlighted that events caused by inattention were most likely to occur due to drivers interacting with wireless devices while driving, followed by other internal distractions and passengers (Neale et al., 2005). Furthermore, drivers who were engaging with a complex visual and/or manual task were three times more at risk of being involved in a crash than drivers who were fully attentive, and two times more likely when engaging with moderately complex tasks (Klauer et al., 2006). This highlights the prevalence of distraction by wireless devices and their potential to impact on the safe control of the vehicle.

Data from a more recent and larger naturalistic driving study, Strategic Highway Research Program (SHRP) 2, further highlights the prevalence of driver distraction. The program, led by the Transportation Research Board, ran from 2006 to 2015 and conducted a large-scale naturalistic study with 3,524 drivers who each participated for 1–2 years across six locations in the USA. Participants' own vehicles were fitted with a data acquisition system that stored data on the forward radar, four video cameras, accelerometers, vehicle network, GPS and on-board computer vision algorithms. All activity across the duration of their participation was stored for analysis. In an analysis of the SHRP 2 to determine the factors relating to crashes, Dingus et al. (2016) found that distraction was evident in 51.98% of normal driving states and 68.3% of the injury/property damage crashes that occurred within the study. Furthermore, it was found that distracted driving results in a crash risk that is two times higher than full attentive driving (Dingus et al., 2016). Dingus et al. (2016) went on to determine that this suggests that 36% of all crashes in the USA could be avoided if distraction did not occur, or 4 million of the total 11 million crashes. A breakdown of the sources of distraction highlighted the heightened risks of handheld devices which increased the risk of incident to 3.6 times higher than attentive driving, and interacting with in-vehicle devices, other than the radio, which had a risk factor of 4.6 higher than attentive driving, thus motivating the need to minimise the driver's interaction with in-vehicle technology. Key ways in which this is currently being achieved is through legislation and the use of design guidelines to minimise the impact of technologies on road safety.

2.2.1.2 Legislation

Legislation on the use of devices and inbuilt technology in vehicles often has to play catch-up, advising on usability after their widespread use. This is notable when discussing the use of handheld mobile phones whose introduction, and increasingly prevalent use by drivers, challenged road safety to the extent that it is now banned by law across many countries in Europe, some states of the USA, Australia, New Zealand, Japan and India, among others. Yet, there are questions being asked on the effectiveness of legislation in resolving the issue. For example the UK banned the use of mobile phones by drivers in 2003, yet over a decade later, the issue of driver distraction remains an issue for road safety management with 1.6% of drivers across England and Scotland observed to be on their phones in 2014 (Department for Transport, 2017). As there were 45.5 million active driving licences on record at the time of this assessment (DVLA, 2015), this represents a large number of drivers who are continuing to use their phones illegally whilst driving. Rather than seeking to explore why drivers continued to use their mobile phones despite the legal consequences for doing so, the UK government has increased the penalties drivers' face if they are caught using their phone while driving. The penalties were initially increased in 2007 from £30 to £60 and again in 2013 to £100. From March 2017, the penalties drivers received for using a mobile phone doubled again to 6 penalty points on their licence and a £200 fine. In the UK, 12 penalty points result in an automatic ban from driving for 12 months. A recent report in the media (from data on phone use since the new legislation became effective through the Press Association under the Freedom of Information Act) stated that nearly 6,000 drivers were caught using their devices in the four weeks post the new legislation change (BBC, 2017).

Mobile phones are not, however, the only technology that drivers have available to them. The availability of functions built into infotainment systems in the vehicle have become incrementally more advanced over time, inviting the driver to perform more complex interactions with vehicle interfaces (Harvey & Stanton, 2013). Yet, legislation has not kept pace with this, and other technologies are entering the vehicle that are not banned by law, such as MP3 players (Lee et al., 2012), navigation systems (Tsimhoni et al., 2004) and hands-free communication devices (Horrey & Wickens, 2002), despite evidence to suggest that they do impair the drivers' attention towards the road. Furthermore, as technology continues to advance with the advent of wearable devices, this imposes an increased threat of driver distraction (e.g. Sawyer et al., 2014). The use of such devices will pose many challenges to legislation and regulation due to their less conspicuous yet more intrusive nature.

Driver Distraction and Sociotechnical System 15

The development of technologies such as navigation systems may provide an improvement from the original methods used to complete the task, that is paper maps (Srinivasan & Jovanis, 1997). Yet, the implementation of these devices including their design and functionality should be carefully considered so as not to lead to any negative or unanticipated effects. For instance, the ability to enter a destination while driving has been found to be a complex visual/manual task (Tijerina et al., 1998; Tsmhoni et al., 2004; Harvey & Stanton, 2013) that was not previously possible with original navigation methods such as reading a map.

Legislation against the use of such devices is currently covered under general distraction laws that require drivers to remain attentive to the driving scene for the duration of the time spent driving. Yet, a definition of what constitutes an attentive driver is not defined. The issues surrounding this will be the focus of Chapter 5 of this book.

2.2.1.3 Design Guidelines

Another way of minimising distraction from technological sources is through the design of the device interfaces to limit the attentional demand that they require or the attentional resources that conflict with the driving task. The National Highway Traffic Safety Administration (NHTSA) set out guidelines in 2013 with the aim to outline a standard for the assessment and measurement of interfaces in the vehicle to minimise their potential for distraction. While other attempts have previously been made to set key principals in the design of in-vehicle devices (e.g. Green, 1999; Society of Automotive Engineers, 2002; Japan Automobile Manufacturer Association, 2004; Alliance of Automobile Manufacturers, 2006), NHTSA built on these principals to integrate new developments in research as well as the current and predicted future of technology development. They stated that visual/manual tasks should be assessed for their applicability for use in the vehicle using eye glance testing in driving simulators or occlusion testing. The guidelines stipulate that the task should not require the driver to take their eyes off the road for more than 2 s for 85% of the time that they are driving and that the total eyes-off-the-road time should not be greater than 12 s. It was identified from the 100-car study data that the risk of a crash occurring greatly increases when drivers take their eyes off the road for more than 2 s (Klauer et al., 2006). While the safe limit of the total eyes-off-the-road time was assessed to be 11.3 s, for comparisons to be made, the total time needed to be a multiple of the 2 s, which is why is it rounded to a total of 12 s. Alternatively, when assessing the distractive effects of in-vehicle technology by occlusion testing, which requires the use of occlusion goggles (or equivalent) that limits the time the user can see the task, the total task duration should not exceed 9 s (NHTSA, 2013). The duration of the shutter opening time in the occlusion task is limited to 1.5 s to adjust for the transition time of 0.5 s to take the eyes off the road and direct them towards the task when driving (which participants do not do in the occlusion task). This method does not, however, take into account the cognitive demands of driving such as holding information about the driving task in working memory or the contextual factors that may impact on the ability of the driver to maintain safe driving practises. There are therefore concerns on its use in assessing the distractive effects of technology in the real world.

The NHTSA (2013) guidelines are based on other previous international guidelines including the Japan Automobile Manufacturers Association (JAMA, 2004), the European guidelines (Commission of the European Communities, 2008) and the Alliance guidelines (Alliance of Automobile Manufacturers, 2006). While there are some commonalities across these guidelines, there are also differences. NHTSA (2013) highlights that the Japanese guidance is more restrictive on the driver, suggesting blocking out functionality while driving to limit the drivers' interaction with devices that may be unsafe to use while driving. They also have more constraining acceptance criteria of 8s total eyes-off-the-road time. By contrast, European guidance allows the driver to choose when they want to engage with the device with no specified acceptance criteria (NHTSA, 2013). The NHTSA (2013) guidelines are based predominantly on the Alliance guidelines that are also less restrictive than the JAMA guidelines in an aim to cover all manufacturers who follow either of the previous guidelines. Thus, there are some differences between manufacturers in their cultural approach to integrating technology within the vehicle.

The implementation of these guidelines is largely positive in creating standardisation in the industry and defining certain limitations of in-vehicle interfaces. Yet, the manufacturers who must abide by them also have a vested interest in providing the driver with competitive technologies that will encourage them to buy the vehicle. Thus, testing to reach levels of performance within the guidelines by manufactures may have the potential to be biased. Furthermore, distracting applications are often preceded by warning that highlights the potential for distraction without actively trying to prevent the driver from engaging with them. This removes liability from the manufactures and transfers it to the driver. In addition to this, the guidelines do not account the context of use within which the technologies are used. Testing with occlusion and other metrics is often conducted in isolation from real-world driving conditions, in driving simulator and laboratories and therefore does not consider the other impacting factors on driver attention in the real world. More holistic views of how drivers interact with technologies and the impact that they have on the maintenance of safe driving performance in relation to all other aspects of the driving task are required.

2.3 THE SOCIOTECHNICAL SYSTEMS APPROACH TO ACCIDENT ANALYSIS

The sociotechnical systems approach is now a dominant approach in accident analysis research (Underwood & Waterson, 2013; Stanton & Harvey, 2017). Dekker (2014) argues that the role of the human has transformed from being the primary actor in the occurrence of adverse events to the recipient of adversity that is created by the wider system. This has become increasingly poignant with the rapid development of technologies that have undergone a transition from fixed entities to facilities that can adapt to individual strengths and limitations (Dekker, 2014). Human needs are now considered to be determined by technological advancement (Postman, 1993; Hettinger et al., 2015). Rather than controlling the individual, the advent of Human Factors research in the 21st century focuses on controlling the technology, the environment and the system that they reside in (Dekker, 2014). Reason (1990) stated that 'human error' was the result of a system that permits (even encourages) certain behaviours that turn out to be erroneous (with the benefit of hindsight). The notion of 'human error' is no longer palatable to many Human Factors researchers (e.g.

Driver Distraction and Sociotechnical System

Dekker, 2002). Instead of determining the human as the root cause of an incident, it is now increasingly the starting point from which to begin investigating accident causation in order to realise the other factors that influence human performance. The organisational and technological factors that contribute to the incident must be recognised so as not to view human behaviour in isolation from its context (Hollnagel et al., 2007). Indeed, it is now suggested that establishing safe systems is the responsibility of those who assess the risks and adverse consequences of the integration of technological advancement and automation within sociotechnical systems (Hettinger et al., 2015). Sociotechnical systems theory draws on the development of the complex system approach which argues against looking towards individual elements but stresses the value of interactions between multiple elements that comprise a system and the wider environment within which they are located (Von Bertalanffy, 1968). A sociotechnical system is one that has both social-organisational and technical components that can be viewed as interdependent, working towards common goals for joint optimisation (Trist & Emery, 2005; Walker et al., 2008).

The systems approach has been considered in relation to specific areas of concern within road safety, such as situational awareness (Salmon, Stanton & Young, 2012; Stanton et al., 2017), young drivers (Scott-Parker et al., 2015), road freight accidents (Newnam & Goode, 2015) as well as driver distraction (Young & Salmon, 2015; Lansdown et al., 2015; Parnell et al., 2016).

2.3.1 THE SOCIOTECHNICAL SYSTEM APPROACH TO DRIVER DISTRACTION

The application of the sociotechnical systems approach to driver distraction has aligned with the increasing prevalence of technologies in the vehicle that has highlighted the complexity of the interacting actors that influence distraction-related events (Young & Salmon, 2015; Lansdown et al., 2015; Parnell et al., 2016). Yet, while the benefits have been realised, this has yet to transcend into effective systemic countermeasures (Young & Salmon, 2015; Parnell et al., 2017a).

It has been previously identified in this chapter that there are many challenges in defining driver distraction, capturing the behaviour and targeting it with countermeasures. Largely each of these challenges has focused invariably on the driver, presenting the driver to be the key responsible element attributed to driver distraction events. The developments in technology that require increased interaction with the driver to provide information, communication and entertainment functions are likely only to enhance the burden on the drivers' attention for the road and the responsibility that they hold in maintaining safe control of the vehicle. Yet, it is the view of the authors that this highlights the importance of taking a sociotechnical systems approach which does not seek to study driver behaviour in isolation but accounts for the wider system within which the drivers' behaviour occurs.

The current proposed manufacturer guidelines for preventing distraction and distraction-related legislation aim to minimise distraction from in-vehicle technology that are still focused on the driver and their interaction with the devices. Furthermore, they strive to identify the cause/effect relationship between the driver and incidents. Hettinger et al. (2015) state that technology is evolving and being implemented into sociotechnical systems simply because it is novel and progressive, rather than due to the necessity or utility that it may have. Furthermore, those who develop and implement such technologies are not best placed to assess the impact they may have on society (Hettinger et al., 2015). It is therefore important to fully recognise the role of technologies within sociotechnical systems, such as that of road transport, and how they may be adversely impacting the safety of the system in order to balance this with the benefits that they may provide. As increasing levels of death by dangerous driving convictions align with the increased engagement with technology in vehicles (Office for National Statistics, 2013; RAC Report on Motoring, 2013), the safety issues needed to be realised.

It is evident that technologies offer desirable functionalities such as navigational information, hands-free communication and even music players to the driver. Yet, balancing the benefits of these devices with their potential adverse impact on the drivers' safe control of the vehicle need to be reviewed. It is the purpose of this book to review the functionalities of these technologies and their interaction with the driver with respect to the wider sociotechnical system in which they are present to determine how distraction from technological sources may be better managed by the system as a whole. The sociotechnical systems approach seeks to explore the social aspects of people and their interaction with technology in relation to the organisational context in which these interactions occur. Therefore, the sociotechnical factors that contribute to incidents resulting from distraction-related events should be recognised to further understanding of how distraction occurs and who is responsible. For example, this will involve determining not just why the driver interacts with technology to become distracted but why device manufacturers are allowed to facilitate distracting technologies in the vehicle to allow the driver to become distracted. As the sociotechnical systems approach to driver distraction has been little considered in the past, this book will seek to determine the relevance, utility and importance of applying the approach going forward. An initial step will be to determine the methodologies through which the sociotechnical system surrounding driver distraction can be studied. This is the focus of the next chapter.

2.4 CONCLUSION

This chapter has provided an initial introduction to driver distraction and the key challenges to the study of the behaviour. It has highlighted the importance of studying the role of technological sources of distraction that have become increasingly prevalent in the vehicle. Whilst technological advancement has some benefits to the development of the automobile, it is highlighted that we must seek to control the implementation of technology that can pose as a source of distraction and therefore adversely affect the safe control of the vehicle. The importance of utilising the sociotechnical systems approach to review this issue is stated. The remaining chapters of this book will seek to explore the possibilities of taking a sociotechnical systems approach to driver distraction. This will continue in the next chapter with a review of the methodologies that are used to study driver distraction, looking at the traditional methods that have been applied in the past and the possibilities of applying systems-based methods which can enable the wider systemic factors to be revealed.

3 Driver Distraction Methodology

3.1 INTRODUCTION

When seeking to explore human behaviour, the methods by which the data on behaviour is collected, analysed and assessed determine, to some extent at least, the type and form of insights that are available to the researcher. Therefore, reviewing the methods that are available to study driver behaviour is a good starting point when striving to obtain the current understanding of the underlying principals and facets of that behaviour. This chapter seeks to provide a review of the methods that can be used to study driver distraction as Human Factors researchers have used a range of methods and measures to study the behaviour.

Within the Human Factors domain, it has been suggested that, despite a change in perspective towards accident causation, the application of methodological tools to assess accidents have been found lacking (Leveson, 2011; Salmon, Goode, Spiertz et al., 2017; Shorrock & Williams, 2016), with a research-practice gap linked to impeding the benefits of a sociotechnical system approach to safety (Underwood & Waterson, 2013). The focus on regulatory guidelines and testing procedures is typically limited to the individual driver and their interactions with in-vehicle interfaces. As noted in the previous chapter, the NHTSA (2013) guidelines outline the recommended procedures for testing the distractive potential of devices in vehicles, including conducting simulator studies using eye-tracking metrics as well as occlusion testing. Sample size and experimental procedure are also outlined. Yet, these methods are inherently focused on the individual, with objective measurements of the driver limited to their interaction with devices under experimental conditions. Whilst the guidelines encouraging manufactures to adhere to a universal set of standards that limit the potential for distractive devices are to be applauded, the wider implications of the sociotechnical system are not considered in the assimilation and application of these guidelines.

Systems thinking has been integrated into the methods used to assess accidents within other domains, such as aviation (Shappell & Wiegmann, 2012), yet its role in providing effective countermeasures to driver error has lagged behind (Stanton & Salmon, 2009). Although the benefits it may have in reviewing error causation and developing effective countermeasures are beginning to be recognised (Stanton & Salmon, 2009; Salmon, McClure & Stanton, 2012; Young & Salmon, 2012; Hughes et al., 2015), the application of these to effective road safety measures is limited.

3.1.1 DRIVER DISTRACTION METHODOLOGICAL CHALLENGES

When deciding upon a method to study behaviour, researchers must determine their specific research questions, the practical and ethical issues that may occur and how they wish to analyse and report their findings (Stanton et al., 2013). Research has targeted multiple sources of distraction, including those external to the vehicle (Dukic et al., 2013), mind wandering (He et al., 2011) and, with a notably increase in the past two decades, mobile phones (see McCartt et al., 2006 for a review), as well as other technologies (e.g. Lee et al., 2012; Tsimhoni et al., 2004). There are, however, several issues that researchers face when trying to study the phenomenon. These relate in nature to the reason why the area is of interest in the first place, predominately its effect on road safety (Young et al., 2008b). There is also a trade-off between control and validity when choosing the method of study (Burnett, 2009). Undertaking study of the driver behaviour in naturalistic settings ensures high levels of confidence that the data captures real-world behaviour, but it also has potential safety implications that may conflict with the ethical governance and acceptable assessment of risk. Driving simulators, however, provide a safe and highly controlled environment from which to examine distraction and the drivers' response to it, but there is less ecological validity as it does not represent the risk and exposure that is found in the real world. Advances in technology have enabled the development of improved simulator fidelity and validity (Young & Lennè, 2017) with full car simulators coupled with simulation software that enables high graphical realism (Kaptein et al., 1996).

Driving simulators have advanced understanding the effects of secondary task engagement on driving. This research has shown reductions in hazard detection (Summala et al., 1998), poor vehicle control (Tsimhoni et al., 2004), attention tunnelling (Reimer, 2009) and slower resumption of control (Eriksson et al., 2017), as well as the compensatory mechanisms employed by drivers such as slowing down and reducing driving performance goals (Alm & Nilsson, 1995). Research has also defined the duration of potentially distracting tasks to be those requiring more than 12 s of attention in total, comprising of no more than individual chunks of 2 s (Klauer et al., 2006; NHTSA, 2013). This quantification of distraction has highlighted the disruption of technologies such as mobile phones that allow the user to engage in calls, messaging, social media and even taking photographs whilst driving and thus justifying the ban on the use of the device while driving in many countries since its widespread use (McCartt et al., 2006). Harvey and Stanton (2013) also found that entering a destination into a satnav whilst driving, a task that is still permitted by many devices, often took longer than the recommended 12 s. Unfortunately, highly controlled experiments in simulators do not provide the context surrounding drivers' engagement with technology, which plays a role in the safety-critical nature of interactions with secondary tasks (Harvey et al., 2011a). It has tended to lead to observations of outcomes once a distractive event has occurred, rather than how the distraction comes about in the first place. Alternative methods are required to gain an understanding for people's experience of distraction, their opinions, knowledge and perception of the behaviour (Ercikan & Roth, 2006). Such approaches are associated with more qualitative and subjective research.

Moreover, as technologies introduce more complexity into the already complex system of road transport (Dekker, 2011; Salmon, McClure & Stanton, 2012), it is important to understand the implications that technological advancement has on the wider sociotechnical system. The driving task is touted to become increasingly 'technology centric' in the future (Salmon, McClure & Stanton, 2012;

Driver Distraction Methodology

Stanton & Salmon, 2009); therefore, methods that are capable of assessing the interaction of multiple actors in the road transport system are required. This includes those actors who influence the design, interaction and presence of the technology in the vehicle. These methods should assess how these actors are enhancing, or hampering, the safe functioning of the road transport system. The systems approach highlights the importance of understanding the shared responsibility for safety, with accidents perceived to emerge from normal behaviour that has evolved over time through interactions with a host of actors, from policymakers down to the surrounding contextual environment (Rasmussen et al., 1997). The theoretical approach applied, and methods used, to study behaviour have important consequences to the way in which the behaviour is viewed and managed (Salmon, McClure & Stanton, 2012; Salmon, Goode, Spiertz et al., 2017).

3.2 CLASSIFICATION OF METHODOLOGIES

The distinction between quantitative and qualitative research is embedded within the social sciences and has contributed to distinct research fields and methodologies (Ercikan & Roth, 2006). Quantitative data can be measured and quantified, whereas qualitative research is more descriptive. Yet, it is important to note that the classification of research as being quantitative or qualitative is not always straightforward (Denzin & Lincoln, 2011). The objective/subjective, dichotomy adds another level to research methodology categorisation (Annett, 2002). Subjective data relates to the individual's personal judgements and opinions, whereas objective data implies impartial measurement of performance (or other metrics). These objective/subjective research foundations feature in both qualitative and quantitative research. The implications of subjective, objective, qualitative or quantitative methodologies that are used to study behaviour are important to realise, as they inform the type of data that can be collected and the knowledge that can be obtained. This is illustrated in Table 3.1 with an example of studying distraction from the task of reading a text message while driving.

The range of methods that have been used to study driver distraction are presented in Table 3.2. The roles that these methods have in the development of knowledge into

---

TABLE 3.1
Research Approaches and the Example Data They Generate

	Quantitative	Qualitative
Objective	Reading a text took the drivers' eyes away from the road for 4 s (e.g. eye-tracking measurement)	Yes, I read text messages on my phone while driving (e.g. tick-box response to a survey)
Subjective	On a scale of 1–10, the driver rated reading a text while driving to be a 7 in terms of its distractive effects (e.g. Likert scale in a questionnaire)	*"I only read a text while driving if my phone is placed in the phone holder, it is switched on to loud mode and I am driving on a quiet road because it grabs my attention"* (e.g. debrief interview)

TABLE 3.2

The Different Methodologies Used to Study Driver Distraction with Examples from the Literature

| | Quantitative | | | | Qualitative | | |
| | *Measuring Behaviours* | | | | *Observing Behaviours* | | |
Method	**Description**	**Example**		**Method**	**Description**	**Example**	
Facilities:							
Driving simulators	Simulators offer a high level of control and manipulation within study design while providing data on the drivers' inputs and a safe environment to observe the effects of distraction. The level of simulation can vary from desktop-based designs to full car simulators that offer higher levels of fidelity	Young and Stanton (2007)	A driving simulator study was used to assess the level of interference between secondary tasks and the primary driving task. Comparing visual and spatial secondary tasks identified the comparative impact on the drivers' attentional resources in the driving task	Naturalistic studies	The drivers' naturalistic behaviour when driving on real roads is captured, usually through video recordings over periods of time. They aim to improve ecological validity by observing the behaviour as it occurs under normal circumstances	Funkhouser and Sayer (2012)	Video recordings of 108 drivers for 6 weeks found drivers to be on the phone for 6.7% of their total driving time
Test-track	Test tracks offer more realistic driving conditions than simulators while also allowing manipulation of secondary task engagement without the safety limitations of driving on actual roads	Ranney et al. (2005)	21 participants drove on a test track while using voice and visual/manual interfaces. The voice system improved peripheral vision while driving but not cognitive attention	Observational studies	Observations of drivers can be taken from the side of the roadside or through photographing drivers to determine the tasks that they are engaging with	Sullman (2012)	Roadside observation of 7,168 drivers found 14.4% of drivers to be distracted

Objective

(*Continued*)

TABLE 3.2 (Continued)

The Different Methodologies Used to Study Driver Distraction with Examples from the Literature

	Quantitative — Measuring Behaviours			Qualitative — Observing Behaviours		
	Method	**Description**	**Example**	**Method**	**Description**	**Example**
Objective	On-road studies	On-road studies can be conducted with instrumented vehicles that are set up to collect a wide range of driving performance metrics whilst also allowing interaction with real-life road conditions	Harbluk, Noy et al. (2007) — 21 drivers drove while interacting with a hands-free device. When performing more difficult tasks, visual attention was focused more centrally and less towards the instruments or traffic lights	Incident reports	The use of incident and crash reporting conducted by police can be used as a database from which to gain an idea of the prevalence of distracting behaviour	Lam (2002) — Distraction accounted for 3.8% of total crashes causing injury. Frequency of crashes decreased with driver age, apart from in the 30- to 39-yrs age group who had a more pronounced incident rate
Measures: Physiological	Physiological	Physiological measures, for example electroencephalogram, electrocardiogram and skin conductance can be used to measure drivers' workload while driving in simulators, test tracks or on the road	Healey and Picard (2005) — 24 drivers drove on the road while their electrocardiogram, electromyogram, skin conductance and respiration were recorded. Their skin conductance and heart rates were correlated to their stress levels	Cross-sectional surveys	Surveys that capture a representative sample of the driving population under investigation can capture the self-reported characteristics of drivers	McEvoy et al. (2006) — 1347 drivers reported lack of concentration, in-vehicle equipment adjustment, external objects and passengers as the most distracting tasks while driving

(Continued)

TABLE 3.2 (*Continued*)

The Different Methodologies Used to Study Driver Distraction with Examples from the Literature

	Quantitative				**Qualitative**		
	Measuring Behaviours				*Observing Behaviours*		
Method	**Description**	**Example**		**Method**	**Description**	**Example**	
Performance	A variety of driving metrics can be recorded relating to the drivers' speed, lateral control and longitudinal control of the vehicle. Driving simulators allow this data to be readily accessible	Strayer and Drew (2004)	20 older and 20 younger drivers' performance in the driving task and in a hands-free phone conversation were compared. Drivers had slower reaction times, increased their following distance and had a reduced speed recovery after breaking	Quasi-experimental	Quasi-experimental design allows drivers to be their own baseline when comparing the effects of interventions, crashes or perceived risk before and after an event	Redelmeier et al. (1997)	Case crossover design with 699 drivers found that use of a phone while driving increased the risk of collision to four times higher than when they were not using a phone while driving. The use of hands-free phones has no safety advantage
Eye tracking	The drivers' visual behaviour is an important metric in driver distraction research to identify where the driver is directing their attention	Sodhi et al. (2002)	Participants drove on the road wearing a head-mounted eye tracker to capture eye position and pupil diameter. When tuning the radio and checking mirrors there were long off-road glances compared to checking the speedometer	Case Studies	Studying a specific event where an accident occurred due to distraction allows the factors leading to the incident to be determined in order for lessons to be learnt	Parnell et al. (2016)	A high-profile case of a driver who hit and killed a cyclist while driving and entering a destination in a satnav was reviewed to reveal the role of systemic actors in the progression of the accident

Objective

(*Continued*)

TABLE 3.2 (Continued)

The Different Methodologies Used to Study Driver Distraction with Examples from the Literature

	Quantitative			Qualitative		
	Measuring Behaviours			Observing Behaviours		
Method	**Description**	**Example**		**Method**	**Description**	**Example**

Objective

Tasks:

Method	Description	Example	(Findings)	Method	Description	Example	(Findings)
Lane change task	Drivers are repeatedly asked to perform lane changes as directed by road signs. Performance in their lane changes is compared to an ideal lane change	Harbluk, Burns et al. (2007)	Lane change initiation and deviation of lane change path was affected by the presence and complexity of a navigation task.	Verbal Protocol	Verbal protocols are used to gain access to the drivers' cognitive processing by asking them to 'think aloud' while driving. It can be used to assess situational awareness and decision making while driving. It should not adversely affect the drivers' performance	Young et al. (2013a)	Participants were asked to drive a vehicle around a test track while they were distracted by the PDT. This was compared to a baseline drive when they were not distracted. Drivers provided verbal protocols in both drives which were used to determine that the nature of errors that were made when drivers were distracted was the same as when they were not distracted; they were just more frequent
PDT	Drivers are presented with randomised visual stimuli at different eccentricities from central line of sight while driving. They are required to detect the stimuli as quickly as possible	Harms and Patten (2003)	Drivers were asked to drive while following a route from memory versus following a navigation system while performing the PDT. Driving behaviour was not effected by condition but PDT performance was reduced when drivers were given visual navigation messages.				

(Continued)

TABLE 3.2 (*Continued*)

The Different Methodologies Used to Study Driver Distraction with Examples from the Literature

	Quantitative				Qualitative		
	Measuring Behaviours				*Observing Behaviours*		
	Method	Description	Example		Method	Description	Example
Objective	Car following	The ability of the driver to follow a vehicle in front through their reaction time, headway and speed allows for the driver's attention and perception of performance to be measured	Alm and Nilsson (1995)	Drivers were asked to follow a vehicle in a driving simulation while using a mobile phone. Their reaction time increased with phone use, but drivers did not increase their headway to compensate for this reduced performance			

(*Continued*)

TABLE 3.2 (*Continued*)

The Different Methodologies Used to Study Driver Distraction with Examples from the Literature

	Quantitative				Qualitative		
	Measuring Behaviours				***Observing Behaviours***		
Method	**Description**	**Example**		**Method**	**Description**	**Example**	
Gap acceptance	Drivers' acceptance for moving into gaps in traffic are used to measure the effect of distractions on the size of gap accepted and the number of collisions	Cooper and Zheng (2002)	Drivers performed the gap detection task while interacting with an in-vehicle phone, providing verbal messages of varying complexity. The use of the phone and verbal message prevented drivers from responding to road surface conditions and limited their performance				

Objective

(*Continued*)

TABLE 3.2 (Continued)
The Different Methodologies Used to Study Driver Distraction with Examples from the Literature

| | Quantitative | | | Qualitative | | |
| | Measuring Behaviours | | | Observing Behaviours | | |
	Method	Description	Example	Method	Description	Example
Objective	Decision to engage with secondary tasks	Drivers' willingness to engage with distracting tasks can be assessed to infer their ability to adapt their interaction with the driving task. Drivers are given the choice to engage with a secondary task at predetermined points of the route. Factors in the situation can be manipulated to assess how they may influence the operational, tactile and strategic level of control	Schömig and Metz (2013) Drivers were asked to drive in a simulator where they were given the choice to engage with a secondary task at certain points along the route. The points reflected critical and non-critical situations. The drivers' decision to engage, and their subsequent performance in the task, were then measured to assess how they integrated the secondary task with the driving task. This showed that drivers were able to interact with the secondary tasks in a situationally aware manner			

(Continued)

TABLE 3.2 (*Continued*)

The Different Methodologies Used to Study Driver Distraction with Examples from the Literature

		Quantitative				Qualitative		
		Measuring Behaviours				*Observing Behaviours*		
	Method	**Description**	**Example**		**Method**	**Description**	**Example**	
Subjective	Surveys	Surveys completed online, in person or by telephone can gain drivers' self-reported engagement in distracting tasks as well as asking them to rate certain behaviours and their perceptions of tasks using Likert scales	White et al. (2004)	Drivers ranked their self-reported behaviour and their perceived risks of a variety of in-vehicle technologies. Mobile phones were perceived as one of the riskiest tasks, yet drivers still reported to engage in the task as they perceive the risk as being greater for others than for themselves	Interviews	Interviews allow a researcher to ask drivers about their driving behaviour and their views on distracting tasks. Transcripts can be recorded and analysed qualitatively through inductive or deductive thematic analysis or with tools such as Leximancer	Musselwhite and Haddad (2010)	Interviews were conducted with 29 older drivers and found that drivers stated their driving skills and attitude were better than when they were young. Yet, they also found some tasks difficult such as reading road signs and maintaining speed. Interviews allowed older drivers to be more aware of their driving limitations
	Video ratings	The use of video clips to represent driving conditions or scenarios can allow drivers to rate their own actions without actually driving themselves	Hancox et al. (2013)	20 participants were asked to rate their willingness to engage with their mobile phone across 15 video clips. Roadway demand and task functionality affected willingness to engage	Focus groups	Focus groups facilitate discussion between a target population on issues surrounding driver distraction to observe social interaction when discussing their behaviour	Lerner (2005)	45 participants were spilt into 6 focus groups of different ages to discuss their motivations, decisions, road situations and errors made when using technologies in the vehicle

driver distraction and mitigation thereof, as well as their advantages and disadvantages, are discussed below. This is by no means intended to be completely comprehensive, but it does include most of those that have proved popular and have shaped the research field. It also shows how the methods fit into the distinct qualitative/quantitative and objective/subjective dichotomies that have already been highlighted, to show how the different methods inform different data outputs. No one method is able to provide insights into all aspects of distractive behaviour; rather, it is advised that multiple methods are required to capture different aspects of the behaviour (Parkes, 1991 in Burnett et al., 2009; Carsten et al., 2013; Hignett et al., 2015).

3.2.1 Objective Quantitative Methods: Measuring Behaviours

The use of objective, quantitative measures is popular in the field of driver distraction as they can quantify specific behaviours in order to determine when distraction occurs and how it can affect driver performance and road safety (Young et al., 2008a). For example, the measurement of driver performance when engaging with new technologies draws on these methods to determine if they are safe to place in the vehicle. This has been used to inform design recommendations (e.g. NHTSA, 2013) and legislation in the case of mobile phones. The use of simulators enables experimental control but may suffer from a lack of realism and limited application to real-world driving conditions (Young et al., 2008a). The use of test tracks offers a balance of both real-world driving and high levels of control, but without the interactions with other road users within the road transport system, so the validity of findings may be reduced. Some studies have been able to research driver engagement with distractions in the real-world settings (see Carsten et al. 2013 for a review). Harbluk, Noy et al. (2007) asked drivers to interact with a road legal hands-free mobile phone device that did not require them to take their visual attention away from the road and recorded their visual behaviour with a head-mounted eye tracker. They conducted the experiment on real roads. While their results had increased ecological validity, due to the on-road setting, and allowed for the inclusion of interactions with other road users, their results may have been affected by the participants being aware they were being studied. The researchers noted that participants *"exhibited considerable safety-relevant changes in their behaviour"* (p. 378; Harbluk, Noy et al., 2007). Thus, the interaction of the complex environmental conditions surrounding the road transport system are difficult to imitate under the controlled conditions required for experimental manipulation (Carsten et al., 2013). It is therefore suggested that on-road methods should be integrated with other methods (Carsten et al., 2013).

New tasks have been developed to simulate the drivers' ability to engage with secondary tasks while driving, such as the peripheral detection task (PDT). Other driving-related tasks, such as car following and gap detection tasks, have been used to assess the drivers' ability whilst they are concurrently engaging with secondary tasks. Although these have become well-used methods, they have limited ecological validity. Tasks given to drivers under highly controlled conditions do not draw upon the same motivations when the driver is, for example, running late for a meeting and needs to call someone to let them know.

Driver Distraction Methodology

3.2.2 Objective Qualitative Methods: Observing Behaviours

Qualitative, objective methods are popular in understanding the magnitude of the issue of driver distraction and the exposure that drivers have to it (e.g. McCartt et al., 2006). These methods offer a solution to the researcher effect as well as providing high ecological validity by measuring the behaviour from afar, using methods to capture large data sets of drivers during their everyday behaviour. This is useful as it can inform on the scale of distraction, for example the estimated number of drivers still found to be using mobile phones despite their ban (McCartt & Geary, 2004). This data can then be used to develop future countermeasures to control the issue. For example, figures corresponding to the number of incidents and fatalities caused by mobile phones are useful data when proposing new bans and increased penalties on the use of the device (e.g. Department for Transport, 2017).

Naturalistic driving studies offer the opportunity to observe the driver during their everyday driving activities and therefore have a high level of ecological validity. The 100-car study was a large-scale naturalistic study that has been run to capture drivers' inattention and its propensity to influence crash risk and other events on the road (Neale et al., 2002; Stutts et al., 2001; Klauer et al., 2006; Dingus et al., 2006). This allowed a large amount of data to be collected on the drivers' engagement with distractions as well as the influence of the environmental context in the natural setting, which is beyond driving simulators. Observations, however, suffer from under-reporting as not all distractions can be easily observed (Sullman, 2012), and social desirability can occur when participants know that they are being recorded. Nonetheless, the recording of the data for an extended period of 12 months in the 100-car study did allow researchers to identify the significant increased risk associated with drivers taking their eyes off the road for more than 2 s (Klauer et al., 2006), which has had important benefits in the design of in-vehicle interfaces (NHTSA, 2013). Alternatively, crash data reported by the police will capture the cause and effects of severe incidents, which is a small subsection of all the distracting incidents (e.g. Goodman et al., 1999). The crash reports are limited by the reporting strategies and the information available at the scene (McCartt et al., 2006).

3.2.3 Subjective Quantitative Methods: Measuring Opinions

Subjective measures are more concerned with the drivers' perceptions of their behaviour and their potential to become distracted. There has been less focus on the drivers' own views of distraction within the literature (Young & Lennè, 2010). Yet, obtaining data on the drivers' perceptions, attitudes and willingness to engage in distracting tasks is important in order to understand the drivers' motivation to behave in a way that may lead to distraction. The use of Likert scales and surveys provides access to the drivers' self-reported perceptions as rankings across predetermined measures, which can then be quantitatively analysed. The identification of the drivers' perception of risk has been linked to their willingness to engage with tasks through such methods (e.g. Nelson et al., 2009). Young and Lennè (2010) showed that drivers are aware of the risks of engaging with mobile phones while driving, yet report that they use them anyway. Access to the drivers' underlying attitudes and perceptions can be targeted by media campaigns and road safety charities to increase risk perceptions and dissuade drivers from engaging in antisocial behaviour. For example, the Department for Transport run campaigns through their THINK! website and road safety resources to establish their road safety message in the UK. The use of ranking scales can be quite prescriptive as they force the driver to comment and rank aspects of their behaviour that are predetermined by the researchers' agenda, rather than concepts they themselves consider to be important (O'Cathain & Thomas, 2004). This may lead to misleading recommendations that may not represent the driver's own perspective.

The Theory of Planned Behaviour (TPB; Ajzen, 1991) has also been applied to the study of driver distraction and intention to engage with distracting devices while driving (e.g. Walsh et al., 2008; Zhou et al., 2009; Chen et al., 2016). This approach has sought to identify the relevance of the components of the TPB to the behaviour. These components include the social norms surrounding the behaviour, perceived control to perform the task and attitude towards the behaviour. The TPB survey can be used to provide rankings on drivers' beliefs about their behaviour across these components. This has led to some contradictions on which components influence the behaviour (Walsh et al., 2008; Zhou et al., 2009), which limits confidence in its findings. It also projects a rather limiting view of the behaviour as it prevents an understanding of other components outside of those comprising the theory (i.e. the wider sociotechnical system).

3.2.4 SUBJECTIVE QUALITATIVE METHODS: OBSERVING OPINIONS

The application of qualitative methods to access the subjective view of the driver through interviews and focus groups offers an opportunity to understand the drivers' motivations and personal beliefs about the nature of potentially distracting tasks while diving, with richer data than come from pre-existing theories of behaviour (e.g. Lerner et al., 2005). The self-report approaches allow drivers to comment on the concepts that they consider being important to distraction, rather than being dictated by the researchers' agenda. As well as enabling an understanding of how distraction occurs from secondary task engagement while driving through the objective measures of distraction, an understanding for why drivers are distracted in the first place can also be addressed. This is a key question for the phenomenon that is generally less pursued (Young & Lennè, 2010). While the use of such methods have been applied to a lesser extent in road safety than the other methods, the use of interviews focused specifically on older drivers and was able to develop an understanding about their perceptions of certain driving tasks (Musselwhite & Haddad, 2010). Musselwhile & Haddad (2010) suggest that the use of such qualitative methods can encourage older drivers to discuss, and become aware of, their driving limitations by enabling them to reflect on their past behaviour. Furthermore, Lerner et al. (2005) conducted focus groups to facilitate the discussion of the use of different technologies by drivers of different ages. An on-road study, run in parallel to the focus groups, allowed drivers to discuss their decision to engage with potentially distracting tasks (Lerner et al., 2005). This method of data collection can inform countermeasures that aim to mitigate the drivers' desire to engage in distractions while

Driver Distraction Methodology

driving (e.g., media campaigns, education, training) and therefore target the source of the issue, rather than focusing on the outcome of distraction incidents which are often reviewed with hindsight (Kircher et al., 2013). Again caution is required to ensure that the views of drivers are not unduly influenced by the presence and views of the researcher. Disclosure of information may include illegal activities, such as using a mobile phone while driving, and so is a sensitive issue. Therefore, anonymity and confidentiality need to be assured.

3.3 SYSTEMS METHODOLOGY AND DRIVER DISTRACTION

Ergonomics researchers have used multiple methods from which to study the phenomenon of driver distraction. Indeed, due to the complexity of the phenomenon, it is recommended that multiple methods are employed to explore all aspects of the behaviour (Carsten et al., 2013). Furthermore, it has also been realised that to fully understand the behaviour, and its relationship to driver error, the contribution of the wider system within which the behaviour occurs needs to be assessed (Stanton & Salmon, 2009; Young & Salmon, 2012, 2015). Young and Salmon (2012) state that the methods that have been used to assess this relationship have focused too heavily on cause/effect relationships between driver performance measures (i.e. the quantitative/objective methods in Table 3.2). Hettinger et al. (2015) highlight that the traditional methodologies that aim to experimentally control confounding variables in order to assess the effect of one or more variables on another is the antithesis of complex system thinking. Complex systems are defined by their interdependence of complexly interacting factors, their adaptability and inclusion of the wider context surrounding the system. Therefore, assessing accidents through the assessment of singular root causes is limiting (Leveson, 2012). The application of these methods in isolation, without accounting for the role of external factors, does not allow for an insight into the impact of the wider system on the behaviour. Assessing external variables, such as the environment complexity which can be manipulated in simulator studies, or integrating methods to review the impact of workload on task engagement across situations, may, however, reveal the contribution of other factors in the development of accidents. Yet, methods that accurately depict errors within the road transport domain have lagged behind other domains (Stanton & Salmon, 2009). This has led to an over-representation of the role of the driver in accident causation, rather than understanding how the conditions surrounding the event may have contributed. This, in turn, is why countermeasures to prevent accidents within the road transport domain have tended to focus on changing the behaviour of the driver, rather than making adjustments to the wider system (Larsson et al., 2010). Focusing on singular sources as the root cause of distraction, such as the driver and the mobile phone, ignores situational circumstances and other actors surrounding the interaction at all system levels, including mobile phone companies, law enforcers and the wider social and cultural context (Young & Salmon, 2015).

A country that is striving to adopt a more enlightened perspective to road safety is Sweden, which has a different approach to driver distraction from mobile phones than most other countries. The Swedish government chose not to adopt a specific ban on handheld devices by drivers, based on research into the effects of phone use on crash risk (see Kircher et al., 2013). Moreover, Sweden argued that enforcing laws to ban mobile phones will likely resolve in the use of alternatives such as hands-free devices which have not been found to be any less distracting than the handheld version (Horrey & Wickens, 2006). The cognitive aspect of conducting a phone conversation means that it still affects the driving task, even when the physical handling of the phone is removed (Alm & Nilsson, 1994; Lamble et al., 1999). Analysis of crash data relating to mobile phone use found that crash risk increased when drivers were using a phone, irrespective if it was hands-free or handheld (McEvoy et al., 2006). Furthermore, Kircher et al. (2013) stated that if a driver cannot phone someone to say they are running late, it may induce stress and deteriorated driving, while having the phone off and being inaccessible may cause drivers to become preoccupied if they feel someone is trying to get hold of them. Sweden's approach to road safety more generally has attracted the focus of many in the research domain (e.g. Elvik, 1999; Rosencrantz et al., 2007; Johansson, 2009; Larsson et al., 2010). Sweden passed a bill on traffic safety in 1997 called 'Vision Zero' that stated their intention to ensure that 'eventually' no one will be killed or seriously injured in the road transport system (Johansson, 2009). This bill strived to pursue alternative avenues to mitigate road accidents that are more in-keeping with modern views of safety management and systems thinking in other domains (e.g. aviation). For example placing responsibility for maintaining road safety across the designers of the system as well as the users of the system. This approach aims to provide solutions to road safety through improving infrastructure and road system design; moving away from the role of 'human error' and a focus on the individual, towards the interaction of multiple factors and designing with the human biological tolerance as well as their limitations in mind (Swedish Road Administration, 2008 cited in Larsson et al., 2010).

While this Swedish approach has already seen reductions in the number of fatality rates attributed to the road transport domain (Johansson, 2009), there are discussions on whether the approach is actually truly systems focused (Hughes et al., 2015). A comparison of the national road safety campaigns in Sweden, the UK, the Netherlands and Australia in Hughes et al. (2015) failed to discern that any of these countries were providing a road safety approach that was truly grounded in the systems thinking. Although road safety practitioners in countries such as Sweden are striving to achieve a systems approach, the interactions between the components that constitute the system are ignored, lacking the insight of interdependence that is integral to a systems approach. Thus, much work is still required to develop the systems approach within the road transport domain.

The development of a sociotechnical systems approach will require methods, or a combination of methods, that assess how the driver interacts with the sociotechnical system, as well as how the sociotechnical system interacts with the driver. While the use of the methodologies detailed in Table 3.2 highlights the drivers' capacity to manage distraction and their exposure to it under specific circumstances, the causal factors and progression towards accidents in the real world is less well understood (Young & Lennè, 2010). Instead, methods, or the combination of methods, that are able to infer the drivers' behaviour and their thought processes in relation to the systems behaviour are required. One example of this is the application of verbal protocol analysis with the visual detection task to assess the drivers' performance, frequency of errors and the classification of these errors when distracted compared to when they are not distracted (Young et al., 2013a). This has revealed that the errors which occurred when drivers were distracted were not only similar to those when they were not distracted, but it has also shown that those errors were more frequent. The verbal protocols also showed the contribution that systemic factors have had in the errors that were observed (Young et al., 2013a).

In addition to the methods discussed previously, sociotechnical systems researchers have developed their own set of methodologies from which to understand accident causation across multiple domains (e.g. Rasmussen, 1997; Shappell & Wiegmann, 2003; Leveson, 2004; Walker et al., 2007; Stanton & Harvey, 2017). While they too have some issues (Salmon, Goode, Spiertz et al., 2017; Underwood & Waterhouse, 2013), they are able to provide an alternative way of assessing safety critical behaviours, including driver distraction (Tingvall et al., 2009; Young et al., 2013b). Young and Salmon (2015) state that there is a *clear role for methodologies underpinned by systems thinking* (p. 358) in driver distraction research and that this is only compounded by the increasing prevalence of technology within the road transport system. Methods that have their foundations in systems theory and have successfully been applied to the transport domain are detailed in Table 3.3.

Young and Salmon (2015) were one of the first research teams to apply systems-based methods to driver distraction with an AcciMap analysis. AcciMap analysis seeks to explore the casual elements and interactions when reviewing incidents by identifying all actors within the sociotechnical system and their interactions with each other in the progression towards incident. The identification of key actors and causal sequences in accident development can be used to inform preventative measures. The initial application of AcciMap analysis to the domain of road safety and distracted driving successfully identified the potential for novel countermeasures to the issue (Young & Salmon, 2015). The use of the hierarchical representation of the road transport domain illustrated that targeting the driver alone is a relatively ineffective way of managing distraction. Elements at the bottom of the system, such as the end user, are poorly placed in the systems hierarchy to establish widespread change across the system (Branford et al., 2009). By contrast, those actors/agents towards the top of the hierarchy are better positioned to facilitate wider reaching mitigation strategies.

Other popular systems-based methods that have been applied to other transport domains include the Systems Theoretic Accident Model and Process (STAMP; Leveson, 2004), and the Human Factors Analysis and Classification System (HFACS; Shappell & Wiegman, 2012). Yet, their application to the road transport domain is sparse. Reinach and Viale (2006) applied the HFACS methodology to the railroad domain to assess accident causation and made modifications to the method in order for it to meet the needs of the domain (see Table 3.3). Other applications of the method have made similar alterations when applying the method, for example for air traffic control (Scarborough & Pounds, 2001) and the military (Olsen & Shorrock, 2010). Although these methods have yet to be applied to driver error in the road transport domain, their application to other domains suggests the potential to do so. This warrants future research to identify the errors incurred by distracted driving and the causal factors that result in such errors in order to identify where countermeasures should be targeted.

TABLE 3.3

Systems Methodologies, with Examples from the Literature

Method	Description		Example	Domain
AcciMap Analysis (Rasmussen, 1997)	AcciMaps were founded within the Risk Management Framework (Rasmussen, 1997) that defines six hierarchical levels of a sociotechnical system that interact with one another to determine the behaviour of the sociotechnical system. AcciMaps have been praised for their ability to generalise across domains while identifying links between actors and contributions to failure that stem from the top of the system (e.g. Government policy and regulators) to the bottom (e.g. end users and the environment). Their graphical representation allows system interactions to be readily visualised to inform novel countermeasures	Young & Salmon (2015)	Young and Salmon (2015) performed an AcciMap analysis to show how countermeasures to distraction affect the levels of the sociotechnical system. This showed how measures such as intelligent lock out systems within portable devices and automated technology detectors can act as preventative measures. The benefits of these countermeasures, over more traditional alternatives that are more reductionist and focus on 'fixing' the driver, are revealed. The method highlighted a range of actors, other than the driver, with responsibility for driver distraction, including (but not limited to) road safety policymakers, media, insurance companies, employers, manufacturers and the in-vehicle devices they design	Driver distraction
HFACS (Shappell & Wiegman, 2012)	HFACS is based on Reason's Swiss cheese model which looks at the interaction between latent conditions (i.e. the inadequate aspects of the system) and the action of the human operator in relation to the accident. The HFACS was originally developed to apply the Swiss Cheese model to the aviation domain, with the addition of taxonomies of failures across the levels of the system; unsafe acts, preconditions for unsafe acts, unsafe supervision and organisational influences. It allows all levels of the system to be considered and investigated in their role in contributing to accidents	Reinach and Viale (2006)	The HFACS was modified to optimise its application to the railroad domain including changing the names of the levels to fit the railroad domain, the addition of a top-level (outside factors) and adding other sub-categories where appropriate. It was then applied to six train accidents to assess the contributing factors. A total of 36 possible contributing factors were identified across the 6 accidents, with active failures and latent conditions apparent in all accidents. Contributing factors included technological environment factors, skill-based errors such as attentional failures, poor resource management and organisational practices amongst others. The analysis allowed for low-cost improvements to be assessed to facilitate immediate benefits to railroad safety	Railroad accidents

(Continued)

TABLE 3.3 (Continued)

Systems Methodologies, with Examples from the Literature

Method	Description		Example	Domain
STAMP (Leveson, 2004)	STAMP is a control-based method that asserts that a sociotechnical system is comprised of hierarchical levels of control where each has safety constraints that influence the system's potential for accidents. Accidents are said to occur due to inappropriate interactions between components across the levels, external disturbances and/or failures in the components themselves	Kazaras et al. (2012)	STAMP was applied to the risk assessment of road tunnels and the road tunnel ventilation system in response to the high number of accidents in tunnels. The STAMP model was able to examine relationships between all elements in the tunnel system, as well as the co-operation of those responsible for tunnel safety. The STAMP model identified the benefits of system thinking in risk assessment, rather than identifying the sequence of events which gives a limited view of accident causation. Alternative methods such as tunnel design were proposed for improving road tunnel safety	Road tunnel safety

3.4 THE WAY FORWARD

In a review of the current ergonomics methods available to researchers and practitioners, Salmon, Goode, Spiertz et al. (2017) stated that *"We require appropriate methodologies that reflect how contemporary models think about accident causation"* (Salmon, Goode, Spiertz et al., 2017; p. 196). This chapter has suggested that the need for appropriate methods and theories that are in keeping with the current trends in understanding accident causation is particularly pertinent to the study of distracted driving, especially with the increasing developments in technology that are both brought-in and built-in to modern vehicles. While the developments in technology offer the potential for enhanced safety, such as driver-monitoring devices (e.g. Gonçalves & Bengler, 2015; Dong et al., 2011; EURO NCAP, 2017), the opportunities for other devices to be used that require the driver to divert their attentional resources away from the main driving goal have been identified (Stevens & Minton, 2001; Stutts et al., 2001; Young & Lenné, 2010; Beanland et al., 2013). Methods that focus on the cause/effect relationship between the driver and distraction have a tendency to focus on the adverse actions of the driver and thus have resulted in recommendations that have predominantly focused on the driver. The application of the sociotechnical systems theory to the study of driver distraction would therefore allow for the wider pool of actors involved in distracted driving to be considered and offer countermeasures which address these actors and their interacting nature in the cause of driver distraction. This will require an understanding of all causal factors that give rise to distracted driving to tackle the issue at the source, rather than once it reaches the end user.

It is advised that future research should move away from the traditional cause-effect models that predominantly attribute road traffic accidents to singular causes such as speeding, inattention, and mobile phone use (Hughes et al., 2015). Insights gleaned from sociotechnical methods that strive to identify the relative impact of all the actors in the system that may influence driver distraction should be reviewed to assess how their actions (or inaction) relate to one another in the emergence of safety or, conversely, accidents. Methodologies, or the creative combination of methods, are required that acknowledge the wider systemic factors surrounding driver distraction. In contrast to the isolation of methods that study the adverse effects of distraction within standardised conditions, methods that can offer the opportunity for identifying the progression towards error in relation to the wider system will facilitate the development of alternative countermeasures that may be more effective than the traditional methods.

Going forward with the work conducted within this book it is recommended that methods of studying the behaviour allow for insights into the social and technological context within which the behaviour is occurring. Inspiration should be drawn from systemic methods such as AcciMaps, HFACS and STAMP, which review the components of incidents in relation to their interdependence and interrelations with each other to provide safety or, conversely, create the conditions for accidents. The essence of such methods seek to determine the wider systemic factors that are involved in the progression towards accident; therefore, the factors involved in the onset of driver distraction will need to be determined. This will be the focus of the next chapter of this book.

3.5 CONCLUSION

This chapter has presented a range of methods that can be, and have been, used to study driver distraction. Each have their own related advantages and disadvantages; it is important, however, to acknowledge that the results are limited by the methodological construction used to elicit them. This chapter has suggested that popular methodologies have focused primarily on the objective measurements of the driver under distracting conditions, as well as striving to understand the magnitude of the issue. The insights from such research have afforded an understanding of the drivers' limited information-processing capacity, which has in-turn informed design standards (e.g., Green, 1999; Klauer et al., 2006; NHTSA, 2013) and highlighted the need for mobile phones to be banned from vehicles (see McCartt et al., 2006 for a review). Yet, as technological advancements increase at a rapid pace, the application of driver-focused research and mitigation techniques that can keep up with new demands on the drivers' attention, and incentivise safety, are required (Salmon et al., 2010; Salmon, McClure & Stanton, 2012; Young & Salmon, 2015; Hughes et al., 2015). Furthermore, the underlying reasons for distracted driving behaviours emerging from the interaction between actors in the road transport system as a whole have been little researched. Instead, cause/effect relationships between the driver and distraction-related incidents have emerged from the application of traditional epistemological approaches to studying the behaviour. The current methods of mitigating driver distraction in countries, such as the UK, predominantly reinforce the individual focused perspective of accident causation. Yet, it is proposed here that the application of methodologies that are able to gain an insight into the wider context surrounding the behaviour, be it from the creative combination of methods or adapted systems methodologies, should seek alternative mitigation strategies which focus responsibility beyond the driver.

The use of such methods will be applied within later chapters of this book in order to obtain a better understanding of the behaviour of driver distraction from technological sources with reference to the wider sociotechnical system within which it occurs. The next chapter will aim to further understand the mechanisms through which driver distraction occurs and how they may be influenced by the behaviour of actors within the sociotechnical system.

4 Exploring the Mechanisms of Driver Distraction
The Development of the PARRC Model

4.1 INTRODUCTION

The previous chapters have identified the possibilities of applying the sociotechnical systems approach to safety in the road transport domain and the potential for the approach to provide alternative countermeasures to driver distraction in response to the rapid development of technology that is now commonly found within the vehicle. It has been identified that, while the relationship between error and driver distraction has been alluded to in the literature, the mechanisms involved are still unclear. To fully recognise the relationship, the causal factors that contribute to distraction are therefore required as well as the impact of the wider sociotechnical system surrounding these factors. Therefore, this chapter seeks to determine what the causal factors of distraction from in-vehicle technology are. A high-profile case study is referred to throughout the chapter to highlight the real-world applications of this research. This will also allow the contextual factors surrounding driver distraction to be determined; these are used to assess the role of the wider sociotechnical system in incidents of driver distraction.

4.1.1 A CASE OF DRIVER DISTRACTION

On a clear Sunday morning in September 2012, Victoria McClure was driving on the A4 near Reading in England when she hit a cyclist, killing him on impact. A subsequent court investigation discovered that 500 m prior to the site of the incident the road was straight which, given the clear visibility of the day, should have allowed Victoria McClure to see the cyclist for approximately 18 s before she hit him if she was driving at the limit of 60 mph, longer if she was driving slower. The court discovered that prior to the incident she had been entering a destination into her satnav, which the court ruled caused her to 'drive blind' for the total 18 s. The lack of skid marks on the road provided evidence that she failed to spot the cyclist, even at the last minute, and perform an emergency brake. In court, the defence argued the cyclist was wearing low visibility clothing that prevented him from being seen. Victoria McClure was convicted of 'dangerous driving' and was sentenced to 18 months imprisonment (BBC News, 2013).

The case of Victoria McClure is an extreme example of the many incidents that occur due to distraction from in-vehicle technology, and it is the focus of much research to actively seek ways to mitigate such events (e.g. Lee et al., 2004; Donmez et al., 2008; Burnett et al., 2004). Indeed, Harvey and Stanton (2013) identified task times for operating different in-vehicle devices and showed that entering a destination into the satnav took longer, on average, than the recommended 12 s maximum recommended interaction time (NHTSA; 2013). This suggests that the task of entering a destination completed by Victoria McClure should not have been approved for use in the vehicle in the first place. This highlights the relevance of 'unruly technologies' within the vehicle that are developing faster than the measures that are employed to respond to their potential safety issues, leaving users exposed to risk (Dekker, 2011).

As noted in Chapter 2, this is not to say that all technologies aim to distract the driver. Driving is a goal-directed behaviour (Summala, 1996; Dogan et al., 2011; Cnossen et al., 2004), with multiple goals that can be run concurrently as well as those which must compete with one another (Dogan et al., 2011). Technology has expanded the potential goals available to drivers, for example making a phone call to the person they are due to meet or driving in an eco-efficient manner in response to eco-displays (Dogan et al., 2011). The adverse effects of telecommunication devices (McCartt et al., 2006), music systems (Mitsopoulos-Rubens et al., 2011), and email systems (Jamson et al., 2004) on driving performance is well documented. However, increasingly, drivers are provided with systems that aim to assist the driver with the driving task, for example Intelligent Transport Systems (ITS), navigation systems and efficiency information systems. Yet, the positive applications of these technologies must be balanced with their potential to implicate the main driving goal, to reach their destination safely (Cnossen et al., 2004). As secondary tasks engage the drivers' eyes, mind and hands away from the road, they reduce the possibility of achieving this goal (Harms & Pattern, 2003; Jensen et al., 2010; Kircher et al., 2014). Thus, a fine line exists between devices that assist the driving goal of arriving safely and devices that detract from it. This line needs to be clearer to drivers, manufacturers and regulators (Cnossen et al., 2004; Dogan et al., 2011).

This research aligns with the view that all tasks that distract the driver from the main driving goal of arriving at their destination safely (Cnossen et al., 2004) are categorised as distracting tasks (e.g. Lee et al., 2008; Pettit et al., 2005; Drews & Strayer, 2008). It views distractions as those that hold the possibility to adversely affect the safety of the road transport system, even if this is not always realised (e.g. Kircher & Ahlstrom, 2016). While this concurs with many other conceptions of driver distraction (e.g. Lee et al., 2008; Young et al., 2008; Regan et al., 2011; Kircher & Ahlstrom, 2016), the importance of the wider system within which the behaviour occurs has also been suggested to play an important role in driver distraction. Yet, this has not been fully realised in the definitions of driver distraction that have been developed to date. This research therefore wishes to determine how distracting tasks that divert attention away from the main driving goal of arriving at the intended destination safely arise, and consequently, how they can be mitigated.

Development of the PARRC Model

4.1.2 What Causes Driver Distraction?

While there has been much research into the principals that constitute distraction (see Chapter 2, Section 2.2), less is known about the process through which distraction arises. Determining how technologies may distract the driver requires an understanding of the mechanisms through which distraction occurs (Young & Salmon, 2012). The multiple definitions of driver distraction are complimented with numerous models that have been developed to understand the behaviour (e.g. Sheriden, 2004; Lee et al., 2008; Wickens, 2002; Hockey, 1997; Fuller, 2000). They each suggest different ways in which driver distraction occurs, and how it may result in distraction-related events that adversely affect road safety.

Previous work had suggested that distraction was simply the result of inattention, that is a failure to look at the critical aspects of the roadway due to reduced awareness (Dingus et al., 2006). Yet, it is now considered that distraction is a subtype of inattention that is distinguished by the diversion of attention away from critical activities to the driving goal (Regan et al., 2011). The Multiple Resource Theory (MRT; Wickens, 2002) suggests that the attentional resources that tasks (both primary and secondary) demand are finite, which limits the ability to engage in more than one task that shares the same demand for resources at any one time. Yet, it has been found that drivers can adapt their attentional resources to facilitate the integration of the secondary task alongside the primary driving task (Harvey & Stanton, 2013). Drivers have also been shown to adapt their driving behaviour to reduce the demand in the secondary driving task by slowing down (Metz et al., 2015) and/or increasing their headway to other vehicles (Jamson et al., 2004). Schömig and Metz (2013) suggest that there are three levels of awareness across which the driver can adapt their interaction with secondary tasks: at the planning level to determine strategies, at the decision level to determine the appropriateness of secondary task engagement and at the control level once the task has been engaged with which covers processes such as slowing down or increasing headway.

The Minimum Required Attention (MiRA) theory (Kircher & Ahlstrom, 2016) is built on the knowledge that the drivers' attentional capacity is fluid and adaptive to the demands of the road environment. It suggests that the drivers' attention in the driving task needs to only be 'good enough' to manage the current demands of the traffic system, thus minimum required attention. This theory is novel as it strives to overcome all previous theories on driver distraction that are subject to the hindsight bias in driver distraction that distraction is only realised once incident or adverse consequences have occurred. It does so by aiming to pre-empt what the required attention is, in order to develop an adequate representation of the driving environment and deem the driver to be inattentive if this level of attention is not achieved. Through determining what is required to allow the driver to be attentive, there is the possibility of classifying and detecting inattention before it occurs (Kircher & Ahlstrom, 2016; Kircher & Ahlstrom, 2018).

Sheridan (2004) reviewed the issue of driver distraction when assessing the behaviour from a control theory perspective. This suggested that driver distraction occurs from a disturbance within the lateral and longitudinal vehicle control loop. This approach highlighted that different outcomes to distraction engagement occur due to contextual differences including the drivers' intention, environment and vehicle response. Lee et al. (2008) state that distraction occurs due to a breakdown in a multilevel control process, the levels of control differing on their time scales. The levels relate to operational control of the vehicles lateral and longitudinal positioning (in seconds), tactile control of the vehicles speed and movement (in minutes) and strategic control of the routes and travel patterns of the vehicle (in minutes to months). The control of the vehicle at each of these levels determines the potential for distraction to occur (Lee et al., 2008). The breakdown is thought to occur due to different mechanisms at each of the levels; resource allocation at the operational level, task timing at the tactile level and goal prioritisation at the strategic level. Conversely, Young et al. (2008) stated that the occurrence of distraction and its impact on driving performance are instead moderated by four factors: driver characteristics, driving task demand, secondary task demand and the drivers' ability to self-regulate. They state that the complex interactions between these factors determine the interference that the secondary, distracting task can have on driving performance.

Overtime, theories of distraction have switched from considering it to be a passive phenomenon that the driver is subjected to, towards a more active process that the driver has control over (Cnossen et al., 2000) and that involves adaption in line with the demands of the road environment (Kircher & Ahlstrom, 2016). Such theories include those that suggest drivers play a more active role in distraction as they structure their performance and adapt to the demands of the task, predominantly by slowing down to increase headway and time to collision (Noy, 1989; Hockey, 1997; Cnossen et al., 2000). Research showing drivers slowing down when on the phone (Rakauskas, et al., 2004) or using navigation aids (Cnossen et al., 2000) supports this. Hockey's (1997) compensatory control theory states that under high demands, drivers adapt their behaviour to prioritise their main task, driving, leaving lesser important goals to decline in order to maintain safety. However, incidents such as Victoria McClure's suggests that drivers do not always prioritise the driving task. It could be proposed that in-vehicle tasks relating to the driving task, for example navigation aids and ITS, are wrongly prioritised as they are perceived by drivers to enhance the main driving goal (Cnossen et al., 2000), when in reality, they are just further distractions. Furthermore, where drivers have a heightened perception of their capabilities while driving, they are likely to wrongly believe they can undertake demanding secondary tasks safely (Horrey et al., 2008). Fuller (2000) suggests under some circumstances drivers may be motivated to take on highly demanding tasks that exceed their capability, even though they are aware of the enhanced risk of collision. The perceived value outweighing the potential risk. This is a key component of the Task Capability Interface (TCI) model (Fuller, 2000). Indeed, many people admit to using mobile phones and other devices while driving despite knowing the associated risks (McCartt et al., 2006; Lerner et al., 2008). Yet, it is unclear what trade-offs drivers make when they decide to engage with technology behind the wheel, what information they are utilising when making their decisions or the causal factors through which distraction occurs (Young & Salmon, 2012).

Furthermore, the factors that have been proposed within the models that have been discussed to date have given little consideration for the role and responsibility of the sociotechnical system to maintain safety and prevent driver distraction from occurring

Development of the PARRC Model

(Young & Salmon, 2015). The models, like the definitions of distraction, are predominantly focused on the role of the driver as the source of distraction-related events. Yet, as discussed in Chapters 2 and 3, the role of the wider sociotechnical system surrounding the behaviour needs to be considered. It is therefore evident that a more cohesive model is required to better understand the issue, the factors involved and the relevance of the sociotechnical system. This can then inform effective countermeasures. The qualities of sociotechnical systems-based models are discussed to determine their utility to the study of driver distraction.

4.1.3 Modelling the Sociotechnical System

When studying sociotechnical systems, there are several key characteristics that need to be considered. A key component of a system is their hierarchical framework that is comprised of multiple levels that interact with one another, such that activity on one level is impacted by those on other levels (Leveson, 2012; Hettinger et al., 2015). The risk management framework (Rasmussen, 1997) details the hierarchical levels that have been applied across many domains. It highlights six hierarchical levels that interact with one another. At the top level, government policies and legislation inform regulators such as the media, manufacturers and road designers. In turn, these influence individual companies, their policies and management which affect the behaviour and environment surrounding workers. This, in turn, influences the workers' interaction with their equipment and environment which reside at the bottom of the system hierarchy. In the vehicle, this bottom level relates to the vehicle and road environment, and the workers are the drivers who must interact with the vehicle in the traffic system (Young & Salmon, 2015). Integral to the sociotechnical system framework are the interactions that occur between the levels, as it is not just the individual elements that are important in maintaining a safe system but their relationships between each other (Cassano-Piche et al., 2009). Policy must propagate down the system to be regulated and adopted by the lower levels. Yet, bottom-up feedback from the lower levels upwards is also important to maintain up-to-date policies on new technological developments (Rasmussen, 1997).

The individual components of the system cannot be studied in isolation, their interdependence is important to recognise (Hughes et al., 2015). Sociotechnical systems are complex systems as their performance emerges from the interactions within and between the components of the system across the hierarchical levels (Hettinger et al., 2015). The interacting elements are thought to maintain an equilibrium through feedback loops of information and control (Leveson, 2004). This allows the system to adapt to maintain effective performance. Safety is a performance outcome of a system; it too is therefore viewed as an emergent property of the interdependencies between hierarchical levels within the system. Consequently, adverse safety incidents can be reviewed through the assessment of the interactions between social and technical factors within the system (Hollnagel, 2009). Incidents are thought to occur within sociotechnical systems due to unforeseen interactions between elements within adverse circumstances that can lead to the emergence of risky circumstances (Perrow, 1984; Leveson, 2012).

Modelling sociotechnical systems can assist in reviewing the systems hierarchy, interacting elements and the possibility for unforeseen interactions to occur (Hettinger et al., 2015). Applying Rasmussen's (1997) framework, Young and Salmon (2015) identified an array of elements that influence distraction-related incidents residing outside of the drivers' control, and thus highlighting the limitations of a driver-centric approach. For novel and influential countermeasures, strategies are required that account for the wider driving system and the interactions between elements within it (Salmon, McClure & Stanton, 2012). No theory or model of distraction from in-vehicle technology has been able to incorporate systems thinking. Yet, model development is advantageous to ergonomic research as models enable predictions of behaviour to be made, as well as being a useful means of presenting and advancing research (Moray, 1999).

This chapter will explore the underlying factors of distraction from in-vehicle technology to uncover how they may relate to an explorative model that incorporates a wider, systems view of the phenomena. This will enable the responsibility that needs to be taken by the system in relation to the factors identified, for the implementation of future alternative and effective countermeasures.

4.2 METHODOLOGICAL APPROACH

To identify the systems factors that contribute to incidents of driver distraction, the complex interrelationships between factors and the causality that exists between them needs to be identified. The explorative approach used to identify the causal factors involved in distraction applied grounded theory methodology to the informative and diverse literature on driver distraction and in-vehicle technology use. This enabled the exploration of the key factors of the phenomena from within the literature it is studied.

4.2.1 GROUNDED THEORY

Grounded theory is not novel to the driving domain. It has been used to explore elderly drivers' behaviour (Musselwhite & Heddad, 2010), as well as their passengers and perceptions of in-vehicle technology (Vrkljan et al., 2007). Furthermore, Stanton and Salmon's (2009) factors of accident causation were founded within the literature on human error in road transport. It is a method used to explore possible issues or phenomenon from within the literature that they originate in, to gain insight into the underlying mechanisms of behaviour (Glaser & Strauss, 1967). It does this by constantly comparing and reviewing data to seek out themes, factors and interconnections in an iterative manner. It is frequently used to develop theories and models without a priori expectations in order to gain a novel understanding of a phenomenon. The advantage of this is that it provides conclusions of high ecological validity from an abundant yet narrow depth of literature across an array of different sources (Glaser, 2001). Its qualitative nature means rich data sets are generated. However, the method is not without critique. It has been suggested that it can oversimplify complex issues and interrelations, as well as constraining analysis such that theory is not guided by data but instead limited by it (Layder, 1998). Despite this,

Development of the PARRC Model

as a method from which to develop a model for further exploration of the current literature, it offers the opportunity for novel insight that builds on current research (Cavaye, 1996). It is also credited for its supplementary use with other methodologies (Strauss & Corbin, 1994).

The advantage of the approach in accounting for wider systems factors within ergonomic research is noted by Rafferty et al. (2010) who applied grounded theory to the military domain to draw key causal factors involved in the safety performance of a system from the literature in which the phenomena was studied. Reflecting on qualitative accounts of the actual behaviour within the field, Rafferty et al. (2010) identified key system factors and the interrelations between them. They also exposed how these factors influenced system performance and the emergence of safety within the sociotechnical system. This enabled novel countermeasures that targeted elements contributing to system safety and factors that led to adversity. It is anticipated that applying the same approach to the road transport system will have similar virtues in understanding how driver distraction influences the emergence of safety from the interconnection of multiple systemic levels.

4.2.1.1 Document Analysis

In order to expose the causal factors involved in distraction from in-vehicle technology, the literature from which the phenomenon is cited was required. It is important that the search criteria for this literature yielded the array of research explicitly looking into distraction from technology. Hence a criterion for included literature was required. The absence of a universal definition of driver distraction was highlighted in Chapter 2, which has implications for reviewing and comparing research (Young et al., 2008; Regan et al., 2011). Yet, commonalities across definitions were evident. This included the notion of 'diverted attention', implicit to distraction terminology (Regan et al., 2011), and a competing activity towards which attention is diverted that consequently affects the drivers' engagement with the primary driving goal of arriving at the destination safely (Cnossen et al., 2004). See Section 2.2 for further discussion on the definitions of driver distraction. These key components of distraction were utilised within the search criteria to determine that the articles reviewed were discussing the same phenomenon of driver distraction. It was also essential that the literature reflected the understanding that driving is a goal-directed activity (Groeger, 2000) with safety as a primary goal and secondary tasks, such as those associated with in-vehicle technology, providing the driver with an increased array of goals, secondary to the primary driving goal.

A comprehensive review of the scientific literature encompassing English-language articles was undertaken using purposeful sampling to select items directly relating to the concepts under investigation. The 'Web of Science' database was used as a tool to locate relevant literature, conducting a combined search over multiple databases linked to the platform. An initial search using the terms 'automobile driving' and 'distraction' revealed 393 items, indicating the extent of the literature on driver distraction to date. Further criteria were required to determine technology-based distractions in line with the definition.

Inclusion criteria: Peer-reviewed research articles were included incorporating journal articles, books and conference proceedings. The original search was filtered by key words. Filtering by 'technology' produced 49 items. Filtering by the term 'goal' revealed 21; however, many of these related to an alternative use of the term by some authors as another word for 'aims' in relation to the research paper. The items revealed by these filters were reviewed in accordance with how they related to concepts under consideration.

Exclusion criteria: Items that were not deemed relevant were excluded, these included papers that did not explicitly state a relation to distraction such as those focusing on substance use or fatigue. All papers that related to distraction from causes other than in-vehicle technology were excluded, for example from passengers, daydreaming, external stimuli or due to inexperience. Furthermore, items that were not thought to represent peer-reviewed research such as policy articles and book summaries were also removed.

Papers meeting the criteria were used to find other relevant papers using cited references. The objectives, motivations, independent and dependant variables and findings of articles were reviewed to determine if they captured distraction from in-vehicle technology as determined by the underlying goals of the driver, the corresponding activities and their relation to safety. In total, 33 citations relating to the management of activities, overarching goals and driver distraction from in-vehicle systems were identified as meeting the criteria. These ranged in publication date from 1986 to 2014, consistent with the time period in-vehicle technology began to develop and became increasingly more common place. It should also be noted that this search was conducted at the beginning of 2015, which is why this is the upper end of the literature obtained in the search.

While 33 may seem a relatively small number of studies to review for the development of a model, it must be considered that utilising goal-based behaviour as a way of analysing distraction is a narrowly focused area. The aim of this research is to develop a model that is grounded in the literature to understand the current factors that may be implicating distraction from this perspective, which has not previously been attempted. The limited literature points to potential gaps in the knowledge base, further motivating the potential to develop a model that can represent the concepts. Lansdown et al. (2015) identified similar issues when reviewing literature surrounding multiple driver distractions from a systems perspective, only identifying 10 papers that were relevant to the search criteria. Their review sought papers that explicitly referenced a systems perspective and multiple sources of attention that led to distraction. These differed to those items sought in this research, which looked for the implications technology had on the goals of the driver and the underlying factors leading to distraction. This research did not require papers that referred to a systems perspective, rather items were utilised to inform on a systems perspective.

4.3 RESULTS AND DISCUSSION

Qualitative analysis of the literature using open coding identified a total of 25 factors as contributing to distraction. The majority of articles had more than one of these factors, indicating potential interactions. Causal factors were ordered by their presence in the literature (based on frequency counts) in descending order and a scree plot highlighted the top five held the largest contribution to the discussion in the literature

Development of the PARRC Model

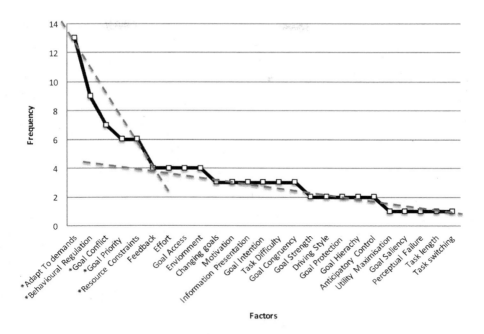

FIGURE 4.1 Scree plot showing the frequency of the factors in the literature. Dashed lines highlight the point of inflexion which identifies the top five factors as those of the greatest importance; these are marked with (*).

(Figure 4.1). These factors were 'adapt to demands', 'goal conflict', 'behavioural regulation', 'goal prioritisation' and 'resource constraints'. Similar comments of circular reasoning can be made as those stated by Rafferty et al. (2010), stating that characteristics may be cited in one instance and then repeatedly analysed in the literature, but it does offer a way of reducing complexity and focusing on a core set of factors as an initial foray. As 25 factors were identified in total, it is evident a large number of factors are discussed widely in the literature.

4.3.1 Causal Factors

The five key factors identified as the most prevalent in the literature are explored in more detail to determine how they provide an insight into the occurrence of distraction. As behaviour is heavily linked to context, the contextual factors that may implicate the outcome of events are also discussed with examples to demonstrate how adverse effects of distraction are linked to each mechanism.

4.3.1.1 Factor 1: Adapt to Demands
Definition: "Adaptation refers to behavioural changes aimed to protect the main task goal in high task demand situations" (Cnossen et al., 2004: p. 219).

When faced with high mental load and complex situations drivers can adjust their behaviour in order to manage the situation. When interacting with a secondary task this often means slowing down, increasing time headway (Hosking et al., 2009) and/or reducing performance goals (Brookhuis et al., 1991). This has been shown with mobile phones (Rakauskas et al., 2004), maps (Cnossen et al., 2004) and eco-driving (Young et al., 2011). Failure to adapt means no added safety margin is given when engaging in risky behaviours (Horrey & Simons, 2007).

Example: A driver's commute to work is heavily congested, but they believe they can spare the attentional resources to reprogram the radio station. The driver may be aware this will heighten their risk of collision, especially in heavy traffic, so they adapt to the increased demand of interacting with the secondary task by decreasing their speed and increasing their headway. Failing to slow down would minimise their time to respond to upcoming events, increasing their risk of collision.

4.3.1.2 Factor 2: Behavioural Regulation

Definition: The self-management of attention, effort, attitudes and emotions to facilitate goal attainment. A goal can be indicative of the behaviour of an individual trying to achieve it, and visa-versa. The ability to regulate behaviour in line with the goals of the individual, the demands of the task and situation is key to maintaining safe driving performance.

Example: If the driver is in the vicinity of a police car they may be more motivated to regulate their driving behaviour in line with safety precautions and legal. For example, abiding speed limits, not talking on the phone, and keeping to the inside lane. When not under surveillance, safety goals may reduce and other goals may counter safe behaviour regulation. When in a hurry, drivers may increase their speed, minimise headway and overtake other drivers to meet the goal of arriving at the destination sooner. Alternatively, boredom on long journeys may increase drivers' engagement with in-vehicle entertainment systems or make phone calls, overriding safety regulation to minimise boredom.

4.3.1.3 Factor 3: Goal Conflict

Definition: The existence of two or more goals that come into competition with each other such that both cannot be completed concurrently without disrupting one another. The degree of conflict depends on the level of shared resources required by the goals (Wickens, 2002; Salvucci & Taategen, 2008) and how they are interchanged (Monk et al., 2004; Lee, 2014). To resolve conflicting goals, it is necessary to either increase the resources available (Young & Stanton, 2002a), if possible, or inhibit other goals and focus only on those which are attainable (Robert & Hockey, 1997). Inadequate goal selection based on inadequate resources has been linked with distraction (Trick et al., 2004; Patten et al., 2004; Regan et al., 2011).

Example: A driver is running late to a meeting so decides to send a text while driving to inform them they will arrive shortly. Texting will result in a conflict between visual resources as they will need to look at the road and the device as well as physical resources to keep their hands on the wheel and hold the phone. To manage the conflict, the driver can choose not to send the text (inhibit the text goal) or pull over to send the text (inhibit the driving goal) or accept that engagement will mean a reduction in either or both tasks' performance. The environmental context is likely to influence the decision, for example heavy rain and surface water may reduce the willingness to engage both tasks at the same time, whereas long stretches of motorway on a clear day may increase willingness. Legislation and its enforcement will also alter the readiness to engage in a distracting activity, if the driver thinks they are unlikely to get caught, they will engage despite being aware of its illegitimacy and associated risks (Young & Lenné, 2010).

4.3.1.4 Factor 4: Goal Prioritisation

Definition: Ordering the sequence of goals that are attended to, based on their perceived importance.

Different goals are relevant to different situations and therefore the priority of a goal is tied to context (Lee, 2014; Kircher & Alhstrom, 2016). The multiple goals available to drivers cannot be completed simultaneously; they therefore prioritise in accordance with their goal hierarchy (Dogan et al., 2011; Young et al., 2011). It is important that the priorities match the current demands (Lee, 2014).

Example: Employers who require employees to drive often supply company cars with in-built communication systems that enable them to be contactable at all times. An important work-related call from a manager is likely to be determined by the individual as a high priority, motivating them to answer even if the environment is highly demanding. Other goals, including safety, may be demoted or compensated. Less important calls may not hold the same effect over the driver.

4.3.1.5 Factor 5: Resource Constraints

Definition: The limited or restricted access to visual, cognitive, auditory and/or physical resources required to perform the driving task effectively. Resources are finite; successful driver behaviour involves manipulating resources to enable their efficient distribution between tasks and according to the situational demands (Wierwille, 1993; Wickens, 2002; Lee et al., 2008).

Example: Selecting a song from a list on an MP3 player (portable or built in) requires high levels of visual monitoring and manual input to complete (Lee et al., 2012). Engaging in this task constrains the visual and manual resources available in the driving task that are needed to monitor the road and control the steering wheel. Effective timesharing of the visual resource requires successive glances back to the road to capture potential hazards and upcoming events while scanning the list of songs.

4.3.2 INTERCONNECTIONS

Safety arises from the complex interactions within a system (Leveson, 2004); therefore, the interconnections are important in identifying how safety emerges. Applying a similar process to that used by Rafferty et al. (2010) and in accordance with the grounded theory methodology, the literature was again reflected on to assess how the five factors identified above interrelate and safety emerges. Referring back to the 33 papers from which the five factors were acknowledged, interconnections were identified by observing any empirically tested connections made in the literature as well as associations made by authors in relating concepts to one another. The number of links between each factor in the literature was then quantified to identify the strength of links within the literature. Connections between each of the key causal factors allow for a novel model to be developed that can depict how safety is impacted by in-vehicle technology use while driving. These connections are illustrated in Figure 4.2 which presents the Priority, Adapt, Resource, Regulate, Conflict (PARRC) model of distraction. This model, developed from the grounded theory approach, illustrates the key factors relating to distraction from in-vehicle technology and their interrelations as represented in the literature from which it originates.

The PARRC model offers a novel approach to the phenomena, with interconnecting factors used to inform on the potential for distraction across incidents. The interconnections may differ for specific distraction incidents, the complexity of the concept means that no two events are the same, however, a generic formulation of the interconnections, as evident in the literature, is given.

The number of cited interconnections between factors was few, with some papers not showing any links between the five factors. The interconnections should therefore be interpreted with some caution. The limited citation of some connections does, however, lend itself to application of further research to evaluate and provide validation for the relationship between factors. Nonetheless, the following comments can be made.

Consistent with the MRT (Wickens, 2002), the PARRC model captures the requirement of adequate resource allocation for effective performance through the influential directional connections to all other elements in the model, with the exception of goal priority. It may be a result of the strong connection between goal conflict and goal prioritisation that prevents resources constraining the prioritised goal; however, this would become clearer with an extended body of literature.

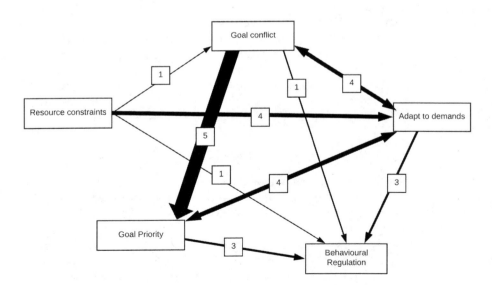

FIGURE 4.2 The PARRC model of distraction from in-vehicle technology. The thickness of the lines suggests the number of the connections made within the literature, as identified in the key.

Development of the PARRC Model

A prominent connection between 'goal conflict' and 'goal priority' is reflective of the need to solve a 'goal conflict' through selecting the goal of highest importance to the individual, situation and/or demands. Without conflict there is no need to prioritise, as all goals would be able to run concurrently with no distractive effects (Wickens, 2002). The PARRC model suggests this through the impacting connection of resource constraints on goal conflict, suggesting that the resource availability limits the goals that may be concurrently engaged. In this respect, it shares the notion of Fullers' (2000) TCI model that capacity must meet demand for performance to succeed, although specifically focusing on capacity as the resources available.

The 'adapt to demands' factor is heavily interconnected within the model, which reflects the extensive literature on the adaption of behaviour in the presence of high task demands (e.g. Noy, 1989; Cnossen et al., 2000; Haigney et al., 2000; Strayer et al., 2003; Schömig & Metz, 2013; Metz et al., 2015; Tivesten & Dozza, 2015). The two-way relationship between 'goal conflict' and 'adapt to demands' is representative of the need to solve conflicts by adapting to demands but also that adaptions can create further conflicts between goals. For example, when slowing down to take a phone call, the demand of the driving task is lowered to facilitate the phone call; however, a conflicting goal to arrive at the destination on time is implicated and the time pressure goal needs readjusting. Thus, caution must be applied not to assume that adapting to meet certain demands resolves all issues; in fact, it can cause more conflicts.

The connection between 'adapt to demands' and 'goal priority' is also consistent with a key concept of Hockeys' (1997) compensatory control theory. Hockey (1997) states resolution of conflicts between goals should prioritise the main task of driving even if this may jeopardise other subsidiary tasks such as those relating to in-vehicle technology. Yet, the literature on driving with in-vehicle technology suggests that often drivers prioritise the secondary task while adapting the driving task, for example taking a phone call from the manager but slowing down and reducing the speed goal (Rakauskas, et al., 2004; Metz et al., 2015). The PARRC model illustrates this through the bi-directional connection between 'goal priority' and 'adapt to demands'. This shows that prioritising the driving task requires adaption but also that adaption to incorporate a secondary task can alter priorities of the primary driving task. Appropriately prioritised tasks prevail without harm to driving performance, for example taking a phone call means slowing down but overall performance does not deteriorate. Yet, where prioritisation is given to a highly valued, but highly demanding task, drivers may be motivated to place themselves at a high level of risk that they cannot adapt against. Hockey (1997) states that under these circumstances, individuals may be able to increase their capacity to manage situations by employing compensatory effort. The relation between resource constraints and adapt to demand in the PARRC model supports the notion that adaption is related to the available resources. Therefore, the model explains the engagement with demanding secondary tasks as an adaption of the capacity to manage high workload that is motivated by effort. The absence of a connection between 'resource constraints' and 'goal priority' could therefore dictate that in order for drivers to prioritise a task in line with the available resources, they must incur some form of adaption through

54   Driver Distraction

task performance goals or effort-related capacity to achieve the goal. The model is therefore somewhat consistent with the notion of compensatory control when faced with resource-demanding situations, whereby a prioritised goal stretches the capability of the driver but they are motivated to engage nonetheless.

The PAARC model also illustrates that salient cues in the environment can cause automatic triggering as they capture the drivers' attention, overriding top-down search strategies (Theeuwes, 2004; Terry et al., 2008) through the direct connection between 'goal conflict' and 'behavioural regulation'. This represents the manner through which the driving task is regulated while it is conflicting with a secondary in-vehicle task for attention. Awareness of environmental cues such as a changing traffic light or a car indicating to pull out have been shown to increase drivers' resumption of the driving task at safety critical points, usually incurring sharp breaking (Jamson et al., 2004). Here no adaption occurs, but neither is it associated with a state of inattention as the driver still attains some level of awareness of the driving task to monitor the environmental cues.

The PARRC model does not therefore suggest distraction results in a failure to attend to the driving tasks but that there are factors in place that allow the driver restructure their behaviour to try and manage the driving task and concurrent technological secondary task. The way in which the factors perform and the interconnections that are realised are specific to each incident. The complexities of distraction must be accounted for by allowing flexibility in the way the model can explain events.

The PARRC model incorporates earlier theories of driving behaviour that highlight passive attentional capacities; the requirement for adequate resources (e.g. Wickens, 2002), and environmental cues that trigger critical events (Theeuwes, 2004). Yet it also incorporates more active factors that relate to the capacity regulated goal engagement of the driver (Fuller, 2000) and their adaption in line with priorities (Hockey, 1997). Application of the PARRC model to a case study was conducted to provide an insight into the impact of systemic actors on these factors.

4.4   APPLICATION OF THE PARRC MODEL: A CASE
STUDY OF DISTRACTED DRIVING

Grounded theory is necessarily explorative in nature; thus in order to perform a preliminary exploration of the validity of the model, its application to a real-world example was conducted. The combined use of bottom-up and top-down testing is a way of assessing the model to determine its efficiency of representing real experiences (Gillham, 2000). Case studies are used frequently across multiple domains within Human Factors research to relate theoretical models and psychological factors to an in-depth analysis of naturalistic behaviour. Examples across domains can be found in aviation (Plant & Stanton, 2012), rail (Stanton & Walker, 2011) and the military (Rafferty et al., 2010). This is especially true of large-scale accidents which, although rare, allow insight into system behaviour and the potential for failure. Single case studies, while limited by generalizability, have been

Development of the PARRC Model

recognised in the generation of theory, definition of concepts and in capturing the context surrounding phenomena (Darke et al., 1998). Furthermore, Hancock et al. (2009) point out the benefits of individual cases in studying human-machine interactions.

To preliminarily explore the potential validity of the PARRC model, the case of Victoria McClure is used (see Section 4.1.1). It is an example of the severe consequences of in-vehicle technology on road safety with a high level of public interest gathered by the press, giving a detailed record of events. The courts portrayal of the 18 s period Victoria McClure took her eyes off the road caused her to serve a prison sentence for 'dangerous driving'. In line with the traditional individualistic approach, the court deemed the driver to be the sole offender, resulting in the heavy conviction. Applying a systems analysis aims to demonstrate wider systemic responsibility and an opportunity for novel countermeasures to prevent such events happening in the future.

Figure 4.3 shows how each mechanism can explain the events in the lead up to the incident and the interconnections that facilitated the behaviour of Victoria McClure. The interconnections suggest how the performance of the system broke down and adverse safety implications emerged, as Victoria McClure engaged in the navigation device for an extended period of time. Some of the system failures reside within the decisions and perceptions of Victoria McClure. For example her failure to recognise the constraints the satnav task would have on her resources when she engaged in the task and established a goal conflict (event a) meant she put herself and other road users at danger as she failed to have enough resources to regulate the driving task efficiently. Furthermore, her failure to reprioritise the driving task (event f), regulating her attention towards the satnav over the 18 s period also support the claims of the court that she was 'driving dangerously'. Yet, applying the systems approach to the PARRC model can highlight failures with the system as a whole that led to the incident.

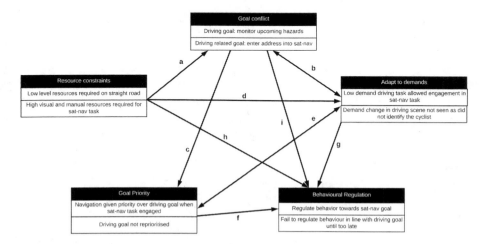

FIGURE 4.3 Application of the PARRC model to the case study with examples illustrating the interconnections in the table.

Event	Description
a	When driving on a straight road spare visual, physical and cognitive resources-facilitated engagement with the satnav goal. The addition of this goal results in a conflict, as resources are taken away from the driving task.
b	Conflict requires adaption of either or both tasks. Demand of road was judged to be easy so driving performance goal was reduced and the secondary task goal was engaged.
c	Conflict between the driving and satnav goal was resolved by giving priority to the satnav goal as determined by the reduced resources required by low driving demands.
d	Visual, physical and cognitive resources were engaged in satnav goal, no attentional resources were directed towards the road; therefore, the driver failed to notice the new demand in the driving scene imposed by the cyclist.
e	Cyclist was not seen so driving task was not reprioritised. Failure to prioritise the driving task and glance back to the road means that the cyclist cannot be adapted to.
f	Failure to reprioritise causes behaviour to be regulated in line with the satnav goal. The driving goal received no visual glances or cognitive interaction.
g	Low driving demand and high satnav demand meant behaviour was adapted to regulate the satnav goal.
h	Visual resources are constrained by the satnav task, behaviour is regulated in line with these constraints to achieve the satnav goal while ignoring the driving goal.
i	Presence of the satnav instates a goal conflict that the driver must regulate their behaviour efficiently to resolve, the satnav goal was achieved but the driving goal was not regulated efficiently.

4.4.1 THE SYSTEMS APPROACH AND THE **PARRC** MODEL

Rasmussen's (1997) risk management theory suggests that elements across six hierarchical levels interact with each other to influence system performance and the emergence of safety. Young and Salmon (2015) highlight the application of each of these levels to driver distraction. They state that the government at the top level puts in place policies and legislation that are then publicised by the media and enforced by police. The regulatory bodies must then influence vehicle manufacturers, road designers and aftermarket device companies to abide the laws before they expose the driver to the in-vehicle equipment and roads they must use. The surrounding environment is included at the bottom level of the system as its changing conditions alter how the driver manages their behaviour in a bottom-up manner, for example water on the road forces drivers to slow down. The basis of Rasmussen's (1997) theory states that elements across these levels have an input into the events that occur. Further exploration of the case study of Victoria McClure considered the potential impacts of wider systemic actors, as initially suggested by Young and Salmon (2015). The outcome of this exploration is illustrated in Figure 4.4 and discussed below.

Development of the PARRC Model

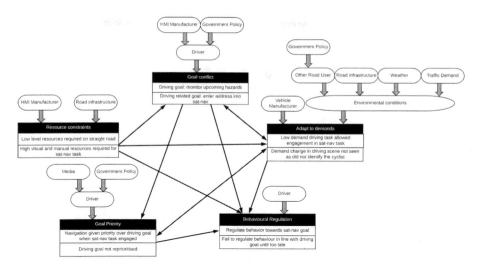

FIGURE 4.4 Application of the PARRC model to the case study showing the wider system elements.

4.4.1.1 Goal Conflict

Reason (2000) stated that adverse events occur due to the continuation of activities in conditions that are undesirable but were permitted by the system. Ultimately, Victoria McClure engaged with the satnav whilst driving because the device permitted her to do so which was determined by the government policy that allows such devices to be road legal and thus conflict with the driving task for the drivers' attention. It was also permitted by the human-machine interface (HMI) manufacturer who chose to design the device for use in motion. Placing the device in conflict with the driving task is thus an initial failure of the system that resulted in the incident.

Satnavs are now a necessity for many drivers and their willingness to engage with them while driving is high (Jensen et al., 2010; Cnossen et al., 2004). They are also argued to have a strong 'point of need' for use while driving (Burnett et al., 2004). Yet, this case study shows that government policy and manufacturers should consider the demands they place on drivers by allowing devices to be in conflict for the drivers' attention while in motion. It is evident that drivers are willing to compromise safety to engage with devices where they are permitted to do so. Some manufacturers have started to prevent the user from interacting with the device while the vehicle is in motion (e.g. Lexus and Toyota). Yet, other manufacturers often allow devices to be driver initiated and in doing so handover the responsibility of engaging with the device while driving to the driver. This is reflective of the different design cultures that are inherent across manufacturers in Japan and Europe (NHTSA, 2013). The liability of the manufacturer is removed by stating guidelines on use, removing themselves from a position of responsibility, even if these are not actively enforced (Sanders & McCormick, 1994).

More should be done across the system to insist the same approach is adopted by all. This requires more top-down control in the form of policy on what is and is not permissible for use when driving, in order to protect the driver and the safety of the transport system. Therefore, a clearer notion of the drivers' capabilities while driving is needed. There are, however, difficulties in achieving this due to issues in measuring visual distraction and stating a criterion for unsafe/safe use (Burnett et al., 2004). The determination of both is highly interlinked to the context of use. This is something that the MiRA theory is striving to overcome by determining the minimum level of attention to be determined attentive and actively monitor this while driving (Kircher & Ahlstrom, 2016).

Workload managers are used by some manufacturers, including those in Japan, for example Toyota whose cultural design ethos is more restrictive of the driver. Workload managers act as a countermeasure to distraction by inhibiting interaction at inappropriate times (see Green, 2004 for a review). Workload is a measure heavily linked to distraction and workload management systems operate by sensing high workload driving situations, such as at intersections, and block any unnecessary and highly demanding tasks, for example, diverting incoming phone calls to an answerphone. Such devices limit the distractive potential of highly demanding secondary tasks (Green, 2004). Yet, manufacturers also seek to meet to the desires of their end users, and research has suggested that drivers do not favour devices that limit functionality, and therefore manufacturers are not motivated to implement such techniques of distraction mitigation (Burnett et al., 2004). Alternatives include driver-state-monitoring technology which seek to monitor the driver's state, alerting them if they show signs of fatigue or inattention. These utilise eye-tracking facilities such as those developed by Smart Eye and Seeing Machines which can monitor the drivers' eye blink frequency, the proportion of time the drivers' eyes are closed and their gaze location relative to the forward windshield. Head movement and positioning can also be used. Such technologies are now a feature implemented by many car manufacturers, for example Toyota, Mercedes and Saab (Dong et al., 2011). They have the potential to restrict the opportunity for distraction, with inattentive states realised by the system and warnings given to direct the drivers' attention back to the road and the driving task.

4.4.1.2 Resource Constraints

The case study illustrates the high visual and manual resource demand of the task 'destination entry', as evidenced in the literature, which led the driver to 'drive blind' for 18 s. In a comparison of time to complete different in-vehicle tasks, Harvey and Stanton (2013) showed that entering an address into a navigation system resulted in the longest task time of all the in-vehicle system tasks tested, with maximum and median response times well above the 12 s recommended limit for in-vehicle tasks (NHTSA; 2013). Such research should be used to inform legislation and manufacturer design against placing highly dangerous tasks into the vehicle to tempt drivers. Manufacturers and developers of devices should design with the users' limitations in mind (Larsson, et al., 2010). Human resources are finite (Wickens 2002; Salvucci & Taatgen, 2008); thus, utilisation of resources other than the visual or manual resources that are critical to the driving task may have allowed visual attention to the road to be minimally effected by the secondary task, for example speech-based input (Ranney et al., 2000).

Another factor impacting the spare resources of the driver is the road infrastructure in the surrounding environment. At the point of the incident in the case study, the driver was on a long straight road which demands a low level of the drivers' visual resources to manage, freeing up resources for the secondary task. In this case, it may have led the driver into a false sense of security and she failed to update her forward view and spot the cyclist. Yet, if a cycle lane had been present, the cyclist may have been better protected from other road users and the driver would be alerted to potential cyclists appearing up ahead enabling them to allocate their resources in line with this, perhaps determining this to be a less appropriate place to interact with the secondary task.

4.4.1.3 Adapt to Demands

As previously mentioned, the straight road lowered the demands of the driving task which was linked with the decision to engage with the device in Figure 4.3 (e.g. events a, b, c and g). There was also high visibility and low traffic demands due to it being a Sunday morning in a rural area, freeing the drivers' resources to engage in the satnav task. Adapting to these low driving demands enabled engagement with the secondary task and consequently reduced the priority of the driving goal, resulting in a failure to monitor and adapt to the cyclist. If the weather conditions had been less clear, or if there was more traffic on the roads, the demand of the driving task would have increased and Victoria McClure may not have chosen to engage with the satnav at that time. Furthermore, in Victoria McClure's defence, the cyclist was believed to have low visibility clothing on that prevented him from being seen. Had the cyclist been wearing brighter clothing he may have been more salient, capturing the attention of Victoria McClure and triggering a reaction. The Highway Code advises the use of highly visible clothing in daylight but there are no laws surrounding this. Indeed, one outcome of this event is the need for increased cyclist awareness on the roads.

4.4.1.4 Goal Priority

The court suggested that the driver was at fault for failing to reprioritise the driving task. Taking a systems approach, Figure 4.4 suggests government education providers and the media may also hold some responsibility for not fully informing the driver of the implications involved in engaging with the navigation system. Fuller's (2000) TCI theory highlights the connection between perceived demands and capability when deciding to engage with tasks. Fully informing drivers of the demands may lessen the potential to interact with navigation systems while driving. The adverse effects of interacting with handheld mobile phones have been widely countered by legislation and widespread media campaigns to increase awareness of the adverse effects of using phones behind the wheel. Conversely, the adverse effects of using a satnav while driving are less well publicised. Satnavs are now commonplace in vehicles, often incorporated into the vehicle design itself, their assistance with the driving task may reduce their perceived distractive potential and heighten perceptions on capability of use. Research has shown the adverse effects on visual/manual interactions with these devices (e.g. Harvey & Stanton, 2013); therefore, education providers and media campaigns have a responsibility to publicise these consequences.

It is important to note that developments of in-vehicle technology are not all negatively associated with safety. Forward collision warning systems are increasingly being installed by manufacturers to warn drivers if they are approaching another road user or object at an unsafe speed and autonomous emergency brake features aim to increase the safety of the driver and road system (Banks & Stanton 2015). Where salient cues in the environment are not recognised, these systems highlight them to the driver so they can reprioritise their goals appropriately. Such a system was not incorporated into Victoria McClure's vehicle, otherwise it would have alerted her to the upcoming cyclist when detected, allowing her to glance up from the road, notice the cyclist and reprioritise the driving task and slow down to pass the cyclist safely. The lack of skid marks on the road evidence that she did not spot the cyclist and perform an emergency brake. The opportunity of vehicle manufacturers to implicate warning systems and autonomous facilities to adapt driving performance to the demands of the roadway is shown in Figure 4.4.

4.4.1.5 Behavioural Regulation

Legislation makes the assumption that driver behaviour is determined solely by the driver and therefore places responsibility on the driver for incidents (Tingvall et al., 2009). Yet, drivers are forced to regulate their behaviour in line with the wider system's actors that are imposed on them. It is evident from this case study that the driver chose to regulate their behaviour in favour of the satnav goal rather than prioritising the driving task to spot the driver, however, as Figure 4.4 identifies, there are an array of factors impacting on this.

Does this therefore suggest that Victoria McClure was not wholly to blame but in fact many other factors in the system failed to protect her and the cyclist? It is hard to determine who is ultimately to blame, nor is it always wise to seek blame (Leveson, 2004), but it is evident that assumptions should not always wholly attribute responsibility to the driver. Exploration of the PARRC model using a case study provides an initial insight into the wider system surrounding driver distraction from in-vehicle technology. In doing so, it highlights responsibility for failures that are attributable to the system as a whole but that are often only linked to the driver.

4.5 GENERAL DISCUSSION

This chapter has applied a grounded theory methodology to establish the key causal factors of distraction that have been discussed in the literature. Through this process, the PARRC model of distraction has been developed. The application of the PARRC model to a case study provided a first step in demonstrating the utility of the model in accounting for the emergence of distraction from systemic actors that may share the responsibility for the adverse consequences of distraction. The development of the PARRC model has both theoretical and practical implications.

4.5.1 THEORETICAL IMPLICATIONS

The causal factors that have been identified build on important aspects of the current literature and thus share commonalties with adaption theories (Lee et al.,

Development of the PARRC Model

2008; Schömig & Metz, 2013) and the role of prioritisation (Hockey, 1997; Cnossen et al., 2000). It highlights the adaptive nature of the drivers' engagement with technological task alongside the secondary task that has been found within the literature. Yet, it also relates to the limited human attention capacity which restricts the number and type of goals that drivers can engage with (Wickens 2002; Theeuwes, 2004). The multiple goals available to drivers require prioritisation, with the relevance of safety fundamental to system success. Cases like Victoria McClure's suggests that the main task goal (arriving at the destination safely) is not always given appropriate priority. The PARRC model presents the causal factors and their complex interdependence which can account for driver distraction from in-vehicle technology. Importantly, it also allows for insights into the role of other elements of the road-transport system. No previous model has been able to account for the impact of wider systems factors in distraction. In order to move away from individual-centred approaches, the role of systemic factors needs to be identified. Utilising the PARRC model, this can be illustrated on a customised case basis.

4.5.2 Practical Implications

Applying the PARRC model to the case of Victoria McClure has initially shown its ability to account for the impact of the wider road transport system on the underlying factors of distraction from in-vehicle technology (Figure 4.4). Although some causes of distraction may be attributed to the driver, such as their desire to engage and their motivation to do so, there is a strong argument to suggest sources outside of the driver may hold some responsibility (Young & Salmon 2015). Elements at the top of Rasmussen's (1997) systems hierarchy, that is government policy and regulations, play a large role in altering perspectives that feed down to the lower levels of the system as policy is adopted. This chapter has highlighted the role that government legislation has on the potential for in-vehicle technologies and portable devices to be used by drivers. Legislation against the use of handheld mobile phones, informed by research on its adverse effects on driving, has impacts throughout the road transport system with media campaigns raising awareness of the risks and manufacturers developing hands-free devices as an alternative. The same is not true of other devices such as navigation aids (Burnett et al., 2004), despite convincing research on their distractive effects (e.g. Lee et al., 2012; Tsimhoni et al., 2004; Horrey & Wickens, 2004; Harvey & Stanton, 2013) and the example case study presented here of Victoria McClure. With more devices being used by drivers, legislation needs to play a more active role in preventing the implementation of technology that has not been fully tested for its distractive effects. It is also realised that the manufacturers of devices also have a role to play in ensuring that their devices are safe to use and anticipate circumstances where this may not be the case. Increasing awareness of the risks devices have on safety and improving design of interfaces to reduce the resources they constrain are some of the potential countermeasure identified in the explorative case study using the PARRC model. These recommendations are fundamentally systems focused with the application of the case study of the PARRC model highlighting how sources outside of the driver contribute to distraction-related events.

4.6 CONCLUSION

This chapter has presented the development of a novel model of driver distraction that has not only identified the mechanism of the behaviour, as they are discussed in the literature, but it has also suggested how the wider system may influence the way in which these causal factors relate to the emergence of distraction-related events. This chapter supports others in the literature who have suggested that distraction from in-vehicle devices cannot be targeted solely with the current individual focused approach (Tingvall et al., 2009; Young & Salmon, 2015). Instead, an approach that unites all systemic elements is called for. This requires furthering our understanding on the causal factors of distractions that have been identified here, as well as the identification of all of the possible actors involved. Further research is required to test the model beyond the case study presented in this chapter to validate the factors and their interconnections that have been identified from the grounded theory approach. This will be the focus of the following chapters of this book.

The application of the PARRC model to a case study has highlighted the impact of high-level policy developers and regulators on safety emergence as well as other systemic factors across the hierarchical levels. The role of legislation, at the top level of the systems hierarchy, on the levels of the system that feed down from this requires further research. This should review the actors that are implicated by legislation and how it influences the mitigation of driver distraction in line with the factors of distraction that have been highlighted here. This will be the focus of the next chapter.

5 What's the Law Got to Do with It? Legislation Regarding In-Vehicle Technology Use and Its Impact on Driver Distraction

5.1 INTRODUCTION

The development of the Prioritise, Adapt, Resource, Regulate, Conflict (PARRC) model in the previous chapter identified five key causal actors of distraction from the literature on driver distraction from in-vehicle technologies; 'goal priority', 'adapt to demands', 'resource constraints', 'behavioural regulation' and 'goal conflict'. The application of the PARRC model to the case study of Victoria McClure, a driver involved in a fatal accident as a result of distraction from a satnav device, revealed the systemic factors that could also hold responsibility for the incident. Using the risk management framework (RMF) (Rasmussen, 1997), the systems actors and their hierarchical levels in relation to each of the five key causal factors in the case study were revealed. Of note was the role of legislation, which is at the top of the RMF and therefore plays an important role in the processes and interrelations between the elements that feed down to the lower levels of the system as policy is adopted. This chapter will focus on the legislation surrounding the use of in-vehicle technologies from a sociotechnical system perspective.

5.1.1 THE ROLE OF LEGISLATION IN THE ROAD TRANSPORT DOMAIN

Since the invention of the modern motor car in the late 1800s and its widespread use from the early 1900s, there have been many developments to the road environment, infrastructure, vehicles, in-vehicle technology, licensure and driver training. Legislation must respond to these changes to ensure safety is maintained alongside new developments. Policies titled "Tomorrows road: safer for everyone" (Department for Transport, 2000) in the UK and 'Vision Zero' (Tingvall & Haworth, 2000) in Sweden infer that legislation is striving to improve road safety in the future. Road safety is threatened by a number of issues, predominately drink-driving, wearing a seatbelt, motorcycle helmets, speeding and driver distraction (WHO, 2016). As has been previously discussed, the issue of driver distraction has become of increasing concern in recent years with the development of technology (Walker et al., 2001; WHO, 2011).

Legislation and regulations must adapt to the rapid developments in technology, yet there is critique that policy change may be somewhat of an afterthought, playing catch-up only after gaps within existing policy have been found (Leveson, 2011). Mobile phones are a key example of this, their use in vehicles was questioned only after risks to road safety were proven (WHO, 2013). Since the early 2000s drivers have been banned from using a mobile phone in many countries such as the UK, Australia, New Zealand, China, Japan, India and EU member states. In the USA and Canada, the laws around handheld mobile phone use varies between states, with 14 states banning their use. Enforcing a ban on specific behaviours aims to target the attitudes of the road users (Chen & Donmez, 2016). Since the mobile phone ban, their use is regarded to have a higher perceived risk than other devices, which is thought to be linked to the increased publicised dangers associated with the ban (Young & Lennè, 2010). Other technological devices that are not banned within legislation and are covered under general laws, using sentiments such as *"You must exercise proper control of your vehicle at all times"* in the UK (The Highway Code, Rule 149) and *"devices may be used as long as it does not detrimentally impact driving behaviour"* in Sweden (Trafikförordning, 1998; Chapter 14.6), are applied. Compared to the definitive ban on handheld mobile phone use, the legal perspective on the use of other technologies is less conclusive both to those who must follow it and those who must enforce it (e.g. Young & Lennè, 2010). Therefore, there is a distinction within legislation which permits drivers to have different attitudes towards devices that are not banned to the same degree that mobile phones are. Yet, research has found other technologies to be no safer than mobile phones (e.g. Horrey & Wickens, 2006; Tsimhoni et al., 2004; Sawyer et al., 2014).

To enforce traffic safety laws, drivers are given penalties in the form of fines and points on their licence when they are found to be contravening the law. In a bid to clamp down on mobile phone use while driving in the UK, the Department for Transport increased the current fine of £100, to £200 and the points on the licence to six points in March 2017, with the hope to deter drivers (Department for Transport, 2016a). These techniques, however, descend from a traditional or 'old view' (Reason, 2000; Dekker, 2002) of accident causation, viewing the driver as unreliable and the main threat to safety (Larsson et al., 2010). Contemporary research favours the 'new' systems approach (Dekker, 2002; Reason, 2000), which considers accident causation to be a consequence of the interrelationships within the sociotechnical system (Larsson et al., 2010; Leveson, 2011; Salmon, McClure & Stanton, 2012; Lansdown et al., 2015). The work in this research joins others in the domain that considers the application of systems thinking to road safety management is a key next step in reducing accidents (Salmon, McClure & Stanton, 2012; Lansdown et al., 2015; Parnell et al., 2016), including those from driver distraction (Young & Salmon, 2015).

This chapter aims to evaluate the current legislation in the UK surrounding the use of mobile phones and other in-vehicle technologies to gain an insight into its efficacy in targeting distraction and maintaining safety with respect to the

Legislation on In-Vehicle Technology

road transport system as a whole. While the UK is focused upon, parallels are drawn from many other countries that similarly ban handheld mobile phones yet have more relaxed laws on other device use in the vehicle. The analysis will pay particular attention to the systems elements involved in the use of mobile phone, and other devices, to determine what impact the distinction within legislation has on the wider road transport system and their responsibility for the emergence of distraction.

5.2 METHOD

As identified in the previous chapter, the role of legislation in relation to the hierarchical levels of distraction is required. Chapter 3 identified the different methodologies that are available to the study of driver distraction. This highlighted the use of systems-based methods that seek to identify the role of systemic actors and the functioning of the sociotechnical system as a whole. In particular, the application of the RMF (Rasmussen, 1997) to distraction-related events has enabled insights into systems-based countermeasure using an AcciMap analysis approach (e.g. Young & Salmon, 2015). While the RMF has been reviewed in the previous chapter with respect to the case study of Victoria McClure, the systemic factors identified will be further explored in this chapter to analyse the laws surrounding in-vehicle technology use by drivers within the context of the whole road transport system. In doing so, it will apply the RMF to the system surrounding the use of technologies by drivers and conduct an AcciMap analysis into the role of legislation on driver distraction mitigation. Adaption of the AcciMap methodology aims to determine general behaviour of the system under normal functioning as Trotter et al. (2014) and Salmon, Goode, Taylor et al. (2017) have achieved in other domains. The legal framework of the UK was used for this analysis but comparisons to other countries legislation are made.

5.2.1 APPLICATION OF THE RISK MANAGEMENT FRAMEWORK TO IN-VEHICLE TECHNOLOGY USE

Rasmussen's RMF (1997) has been applied across multiple domains that comprise sociotechnical systems such as food safety (Cassano-Piche, et al., 2009), public health (Vicente & Christoffersen, 2006), outdoor activities (Salmon et al., 2010) and road transport (Scott-Parker et al., 2015; Newnam & Goode, 2015), including driver distraction (Young & Salmon, 2015; Parnell et al., 2016). This framework typically features six hierarchical, cohesive and interactional levels of a system: the government, regulations, company, management, staff and work. A first step in assessing the impact of legislation on the system was to apply the RMF to driver distraction legislation. This is graphically represented in Figure 5.1. The * indicates additional levels included to apply to the road transport domain.

An initial review of the legislation identified that high-level elements outside of the Government are involved in legislation development. The design and development of in-vehicle information systems is incorporated into international principals (e.g. International Organisation for standardisation, ISO) which set a standard for all

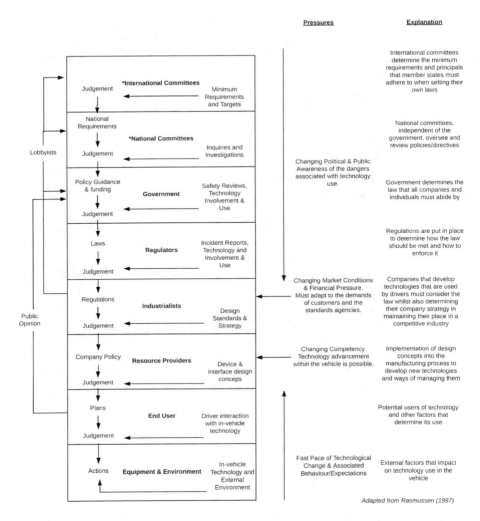

FIGURE 5.1 An extension to Rasmussen's RMF (1997) applied to in-vehicle technology use while driving.

countries conforming to the international committees. National bodies coordinate national standardisation, distributing responsibility across the different governmental departments to enact the ISO within national policy. In the UK, the British Standards Institute (BSI) sets national standards, collectively developed by a technical committee formed of organisations, consumers, industrial bodies, researchers and other experts within the field. These contributors must come to a consensus on the standards required. The traditional six levels of Rasmussen's sociotechnical system have therefore been expanded in Figure 5.1 to include an additional two levels: international and national committees. Governmental departments outline specific policies in line with the international and national standards enacted by relevant government bodies, such as the Department for Transport. Regulators must then control, enforce and inform use related to legislation at the lower levels of the system. Application of the RMF to driver distraction from in-vehicle technology suggests these levels need to be adapted to be more domain specific. The 'Company' level in the original RMF has been re-termed 'Industrialists' to refer to the key industrial organisations related to in-vehicle technologies and their use by drivers in the road context. These industries are comprised of the 'resource providers', the next hierarchical level renamed from 'Management'. Resource providers include the departments involved in the development, design and provision of the resources available to drivers. The drivers and all other 'end users' are reflected in the next level down (renamed from "staff"). The end users are distinct from the equipment and the environmental context they engage with, which is the bottom hierarchical level 'Equipment & Environment'. These adapted levels are relevant to all countries.

The public opinion feedback loop is an important part of the RMF, the nature of democratic societies requires the government and policy makers to pay attention to the opinions of their citizens (Weinburger, 1999). However, the extent to which public opinion has the ability to impact on policy is an area of contention (Burstein, 2003). Notably, the desire to engage with new technologies by drivers should not be given the same weight as the potential for traffic fatalities and incidents that may occur as a result. An additional feedback loop between the lobbyists at the industrialist level and those higher up highlights the influence that lobbyists from vehicle companies can have over legislation implementation within this domain.

Legal requirements change with time, and the respective committees must ensure laws are adequate for the current climate. Systems migrate towards accidents through the natural process of adapting to pressures placed upon them (Rasmussen, 1997). Figure 5.1 highlights the pressures evident in the use of in-vehicle technology. At the lower levels of the system, technological advancements develop to meet drivers' needs and desires to remain connected while driving (Walker et al., 2001). In a bottom-up way, these impact on the automotive and technology companies alike, who must facilitate these advancements and driver requirements. As in-vehicle design develops, it has become an increasingly important competitive element for car manufacturers and technology companies. Yet, manufacturers must also adhere to human-machine interface (HMI) design guidelines and government standards which set the top-down pressures on in-vehicle technology design to ensure technological advancements are applied safely (e.g. NHTSA, 2013). Pressures are normal within systems, but over time, they can push the system too far, such that efficiency risks the performance of the system as a whole. Rasmussen labelled this point as the boundary of "*functionally acceptable behaviour*" (Rasmussen, 1997, p. 189). Reviewing distraction from the systems approach hopes to assess how the system may allow the efficiency of engaging with communication, information and entertainment devices while driving without yielding to distraction. To understand the workings of the system under current legislation an AcciMap analysis was performed.

5.2.2 Application of the AcciMap Analysis to In-Vehicle Technology Use

The AcciMap accident analysis methodology is often used within systems theory to graphically represent a sociotechnical system and identify key causal actors/decisions in the route to incident (Rasmussen, 1997). The AcciMap methodology is cited to be the most useful in analysing accidents in a safety critical domain (Salmon, Cornelissen & Trotter, 2012). Yet, initial applications of the method were limited by an inability to analyse general behaviour, or explain the emergence of accidents from normal behaviour (Salmon et al., 2015). Other accident analysis methods apply to multiple events by using taxonomies, for example the Human Factors Analysis and Classification System (HFACS; Shappell & Wiegmann, 2003) and Systems Theoretic Accident Modelling and Processes model (STAMP; Leveson 2004). These taxonomies determine the general failure modes that are used to determine the causal events of accidents. However, HFACS is less convincing in determining accountability in the higher-level factors (Salmon, Cornelissen & Trotter, 2012). The benefit of the AcciMap approach is that it allows for the assessment of high-level factors, and this chapter hopes to show its application to the international context, as has been achieved with the application of STAMP (Salmon et al., 2016). Recent work has suggested that AcciMap analysis can be adapted to infer more generalizable outcomes. Trotter et al. (2014) applied the AcciMap framework to assess improvisation within sociotechnical systems, and successfully evaluated negative and positive outcomes to suggest it is able to assess normal behaviour within events. Furthermore, a recent application of the method within the outdoor activities domain has shown that AcciMaps and the RMF may be utilised in developing a new incident reporting and learning system within this domain, which facilitated an AcciMap analysis of 226 incidents (Salmon, Goode, Taylor et al., 2017).

This chapter proposes the application of the AcciMap methodology to further develop its accountability for normal behaviour in the emergence of accidents and its application outside of singular events. In the same way that other methods use taxonomies to identify classified and general failures, here the general causal factors of distraction from in-vehicle technology are utilised to determine how the actions and decisions made by elements within the system assist in the facilitation, or mitigation, of distraction from in-vehicle technologies. The general factors of distraction from in-vehicle technology, 'goal priority', 'adapt to demands', 'resource constraints', 'behavioural regulation' and 'goal conflict', which comprise the PARRC model (Chapter 4; Parnell et al., 2016), will be used. These five factors determine the underlying characteristics of a system that make it liable to incident from in-vehicle technology (Parnell et al., 2016). The factors are briefly outlined in Table 5.1, see Chapter 4 (Section 4.3.1) for further detail.

The utility of the PARRC model in assessing the factors through which systems elements influence the outcome of distraction from technology use was shown in the case study in Chapter 4 (Section 4.4). It is applied here to the AcciMap methodology to taxonomically classify the actions of actors in response to current legislation. This will allow the AcciMap to classify the actions of each element to causal factors of distraction that are specific to the domain, rather than generalised taxonomies which feature in the STAMP and HFACS methodologies but that have not yet been applied to driver distraction.

TABLE 5.1
Definition of the PARRC Model Factors

PARRC Mechanism of Distraction	Definition
Goal priority	The multiple goals drivers face cannot be completed simultaneously; they need to be prioritised in accordance with goal hierarchy. It is important that the priorities match the current demands to maintain safety.
Adapt to demands	The increased mental and physical demand associated with engaging with secondary tasks while driving requires adaption of either the primary or secondary task, or both.
Resource constraints	Attentional resources are finite; successful driver behaviour involves manipulating resources to enable their efficient distribution between tasks and according to the situational demands.
Behavioural regulation	The self-management of attention, effort, attitudes and emotions to facilitate goal attainment.
Goal conflict	The existence of two or more goals that come into competition with each other such that both cannot be completed concurrently without disrupting one another.

5.2.3 AcciMap Analysis

The first step in AcciMap development is to formulate an actor map, locating all actors comprising the system across Rasmussen's adapted hierarchical levels (Rasmussen, 1997). The formulation of the PARRC model using a grounded theory methodology based within literature on technology distractions is detailed in Chapter 4 (Section 4.2). This literature was reflected on again in developing the AcciMaps. The interrelations between each of the elements within the literature were explored, including the decisions and actions made at each level. Mobile phone use and other technology use were separated and the elements and actions associated with distraction for each were used to populate different AcciMaps. The actor map and AcciMaps were developed before being reviewed and refined by two subject matter experts with approximately 40 years of combined experience in Human Factors Engineering.

In line with standard AcciMap methodology (Stanton et al., 2013), each of the decisions and actions between the interrelating elements was assessed to determine the errors that had been identified, or where potential for error had been highlighted. Utilising the knowledge of the two subject matter experts, the relation of each of the systems elements across the RMF levels to the wider systems issues were determined. This then enabled the classification of systemic actions and interactions to the factors of distraction highlighted in the PARRC model. This was achieved in the same way as in the case study example in the development of the PARRC model in Chapter 4 (Section 4.4).

Here it was highlighted that some of the systems elements were partaking in actions and decisions that were actively seeking to reduce distraction through the PARRC factors, while others were leaving the system vulnerable through failing to take account of PARRC factors. For example the case of the media and their

70   Driver Distraction

campaigns (e.g. THINK) that aimed to target the drivers' attitudes towards different driving behaviours were found to heavily target the use of mobile phones by drivers. By highlighting the fact that the driving task is more important than the phone tasks, they target the goal priorities of the driver. Therefore, within the mobile phone AcciMap, the media element was deemed to be supporting the goal priority factor in mitigating distraction. However, the use of other technologies while driving has not attracted the same level of media attention; the media are therefore not setting the drivers goal priorities to the same degree within the other technologies' AcciMap, leaving the system vulnerable to distraction via this mechanism. A further example to illustrate the development process of the AcciMap is that of road safety charities. Road safety charities are also regulators of legislation. Yet, road safety charities in the UK do provide information relating to the adverse effects of mobile phones as well as other technologies; the goal priorities of the driver are therefore targeted and supported by road safety charities in both AcciMaps. Distinctions between the ability of the systems elements to support the system in preventing distraction were sought and highlighted within the AcciMaps through highlighting the relevant factors to distraction for each systemic element. The mechanism is shown in black where the element is supporting the system and in grey where actions by the systems element are deemed to be leaving the system open to distraction via the PARRC factor. Each element was taken in turn and the process of classifying its interrelations with the other elements was undertaken to assess its impacts on the PARRC factors. Both the actor map and AcciMaps were developed and then reviewed and refined by the two subject matter experts.

5.3 RESULTS AND DISCUSSION

Development of an actor map found a total of 40 possible actors, spanning eight hierarchical levels relating to in-vehicle technology use (see Figure 5.2). This highlights an array of factors, beyond the drivers' control, that must account for some level of responsibility in the maintenance of safety within the system.

The distinction between legislation on mobile phones verses other in-vehicle infotainment systems is considered using separate AcciMaps (see Figures 5.3 and 5.4). For clarity, each of the elements is referenced with a letter to assert its level and a number to locate it across the level. The AcciMaps detailed in the following figures can be read both laterally and vertically along the interconnecting arrows between elements. Following the connection across the levels will illustrate how the different elements are connected.

5.3.1 ACCIMAP OF PHONE USE

Figure 5.3 highlights the elements that comprise the system surrounding mobile phone use in vehicles within individual boxes across the eight identified levels of the RMF. The PARRC factors are also presented in each box. The top four levels of the system (international committees, national committees, government and regulations) are shown in Figure 5.3 to be effectively accounting for the PARRC factors of 'goal conflict' and 'resource constraints'. The international bodies state the dangers

Legislation on In-Vehicle Technology

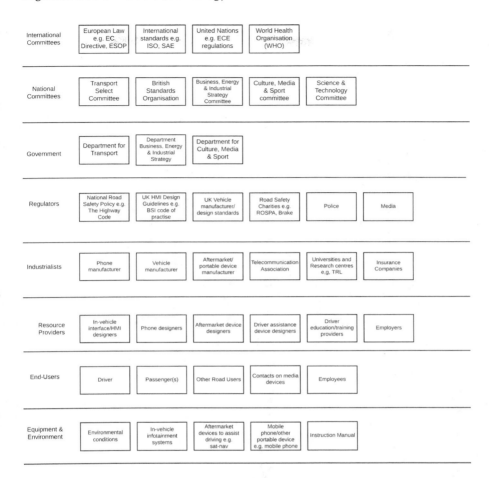

FIGURE 5.2 Actor map containing all actors relevant to the use of in-vehicle technology by drivers.

of using handheld mobile phones, informing on the conflict that arises between the drivers' ability to attend to the road and talk on a handheld phone at the same time (IC2–4). The European Statement of Principals (ESoP) (IC1) suggests that *"Nomadic systems should not be used hand-held or unsecured within the vehicle while driving"* (Commission of the European Communities, 2006, p. 238). This suggests that handheld devices are not recommended for use whilst driving (Commission of the European Communities, 2006). Yet, ESoP goes on to state that mounting the device enables better control and therefore is permitted (when conveniently placed).

National committees inform on the application of the international standards into national law. In the use of mobile phones, committees involved in technology innovation (NC3), industrial strategy (NC4), media (NC5), as well as the transport select committee inform on national requirements. UK committees state handheld phones should be banned. They implement legislation in the UK that prevents them from coming into conflict with the driving task, embodied by government legislation, the

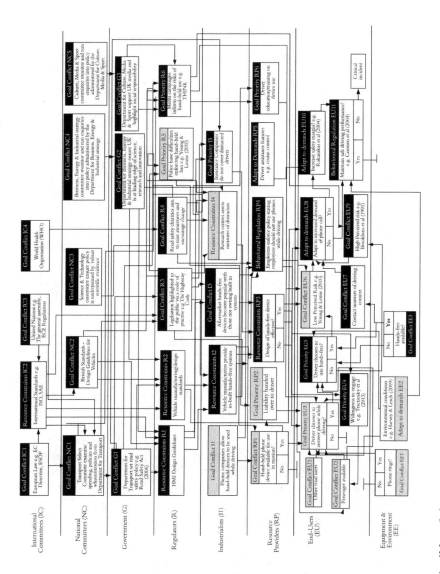

FIGURE 5.3 AcciMap of phone use.

Legislation on In-Vehicle Technology 73

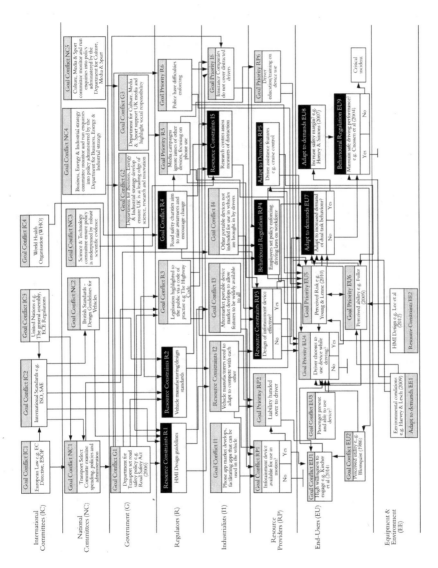

FIGURE 5.4 AcciMap of other in-vehicle technology use.

Road Traffic Act (1988) and the Road Safety Act (2006) which must then be regulated nationally (G1). The Highway Code is a public document which informs all road users of the rules of the road (R3). It allows legislation to be easily accessed and understood by members of the general public in order for them to apply it. Road safety charities also act as regulators by actively informing the public on the laws of the road and how to be safe, raising awareness of the dangers of handheld phone use and the preventative measures against distraction (R4), for example switch off before you drive, do not call others when you know they are driving. Other regulators include HMI guidelines (e.g. NHTSA, 2013) which are accessible to designers and vehicle manufacturers and must be abided by when designing technology accessible to drivers (R1). These guidelines relate to the international and British design standards (IC2, NC2). As handheld phones are banned, HMI designers facilitate communication via hands-free devices. Device manufacturers capitalised on this gap in the market, developing hands-free devices to allow drivers to talk on the phone while driving, which have become very popular (C3). Vehicle manufacturers are pressured into incorporating them into the in-vehicle system design and aftermarket device manufacturers cater for those with older vehicles that do not have the facility built in (C2). These devices must be designed to meet the HMI guidelines and standards that are nationally upheld, for example NHTSA (2013), which limits the length of time it takes to complete a task on a visual manual display while driving to a total of 12 s, comprising glances lasting no longer than 2 s (Klauer et al., 2006; NHTSA, 2013). Yet, although they may be permitted by law, they still have associated risks. Research centres suggests hands-free may be just as distracting as using handheld phones (C6) (Redelmeier & Tibshirani, 1997; Strayer & Johnston, 2001, Horrey & Wickens, 2006). The Highway Code makes reference to their potential to distract but permits their use over handheld options. The responsibility to engage is left to the driver (M2). Yet, the driver has a low perceived risk of the use of hands-free devices compared to the handheld alternative (EU6) (Young & Lennè, 2010). Goal conflict is therefore shown to be unsupported by the system at the end-user level (EU1-11).

At the company level, actors must abide by the regulations and enforce their own company policies to ensure they are meeting the regulations (C1–7). There is focus on making the use of mobile phones an undesirable option, with media campaigns focusing their attention on raising the profile of the risks of mobile phones and insurance companies not covering incidents where a phone was in use (C4 and C7). This ensures that drivers take responsibility for their actions if they decide to use handheld phones while driving, although there are issues proving that drivers are on their phones at the time of incident and often drivers believe that they can get away with using their phone without getting caught (Young & Lenné, 2010). Indeed, without patrolling the roads on the lookout for handheld phone users, the police have a difficult job in enforcing the laws (C6). Furthermore, phone companies allow phones to be used inside vehicles and therefore enable them to be in conflict with the driving task (C1). Stopping the user from engaging with their device is obviously not the desired practice that mobile phone companies want to employ. Yet, their inclusion in the AcciMap in Figure 5.3 shows that they have some responsibility for distraction incidents related to handheld phone use. There are some applications available for smart phones which can limit the usability of the phone while driving and monitor performance to increase safety, but

Legislation on In-Vehicle Technology

the uptake of these applications is a complex issue (Kervick et al., 2015). The phone company is thus handing over responsibility to the driver, allowing interaction to be determined by the situational and motivational factors of the end user as they prioritise their goals. Yet, preventing the responsibility from reaching the end user would stop other factors from influencing the use of the phone while driving. Phone companies are therefore shown not to support goal conflict in Figure 5.3 (C1). The role passes to driver education providers to manage the priorities of the driver and enforce legislation on the dangers associated with using phones while driving at the training stage (M6). Yet, this is not enforced after training is completed and drivers receive their licence. Another management tool recently developed is driver assistance features, such as advanced warning systems and hazard detection functionalities which aim to enhance the drivers' safety and add an additional line of defence to alert the drivers' attention back to the road when they may be distracted (M5). Yet, these systems may disturb the driver' risk perception (EU6) (Robert & Hockey, 1997). Instead of enhancing the drivers' safety, warning systems can enhance the drivers' perception of safety leading them to feel protected and therefore increasing the potential to engage in riskier behaviours (Dulisse, 1997). Thus, the driver is ultimately left to manage the risks placed in their environment when they are poorly advised on the actual scale of the risks.

Nonetheless, when drivers are given the responsibility to manage their own interactions with their phones while driving, there is a high level of responsibility that the sociotechnical system is placing on the driver. The decision to engage is hard to control without preventing it from occurring at higher levels. The priority that the driver assigns the phone call is also subject to a number of factors which determine the drivers' motivation to engage (EU4). The unpredictability of this decision means that it is a vulnerable aspect of the system and is highlighted grey. The presence of other road users in the system and passengers, actors found to influence the decision, are equally random (EU1 and EU2). Prohibiting phones from vehicles would eliminate this variability and facilitate goal prioritisation away from distracting activities. Instead the system is left vulnerable to the interactions of the end users and their interactions within their environment (EE2).

For the most part, Figure 5.3 highlights that legislation at the top level of the system that defines the ban on handheld phones is effective in motivating the other levels of the system to provide support against distraction from mobile phone devices. However, caution is needed to prevent drivers from being directed towards alternatives that are not as safe as they may be perceived, such as hands-free devices.

5.3.2 AcciMap of Other Technology Use

Figure 5.4 includes significantly more grey boxes across all levels of the system than the AcciMap corresponding to phone use (Figure 5.3), inferring that less of the PARRC factors are deemed to be supported by actors under this legislation. This includes elements at the very top of the system, the international committees and regulatory bodies. Unlike Figure 5.3, where the international laws highlight the dangers of phone use, the suggestions made for other technologies are less conclusive (IC1–4). Although they are aware of the potential for distraction from a range of technologies, quite how they may conflict with the driving task is less well defined. International standards apply criteria to in-vehicle system design (e.g. NHTSA, 2013). This feeds down through the national committees to the government who state legislation on devices that impact the ability to control the vehicle, stating drivers must not be *"in a position which does not give proper control or a full view of the road and traffic ahead"* (Road Traffic Act, 1988, Section 2, 41D) (NC1 and G1). Yet, they are less specific on how to avoid distraction and exactly what proper control means. As is evident in Figure 5.4, there is a clear lack of awareness for the type of goals that technologies pose as conflicting with the drivers' attention for the road. This is embodied at the regulatory level, with the Highway Code stating less clear information on how to avoid distraction from in-vehicle technology (the Highway Code) (R3). Some regulation is evident; road safety charities aim to target all aspects of safety and therefore pay attention to the use of other devices behind the wheel, providing leaflets and information on their distractive effect (R4). Yet, these are only available to those willing to look for it and are notably less widely available compared to those corresponding to phone use.

The resource constraints of the driver are accounted for in the system surrounding in-vehicle technology. As with phone use, research into the effects of driving with a secondary task has shown the drivers' attentional capacity while engaged in the driving task (C7) (see Young et al., 2007 for a review). This informs the guidelines put in place for HMI designs and vehicle manufacturers to follow when designing driver interfaces (R1 and R2) in line with the international and national standards (IC2 and NC2). Yet, as more companies begin to enter into the in-vehicle technology domain, the exact resources that are constraining the drivers' attention are less well known (C1–4). The need to compete and remain profitable means that vehicle and device companies need to offer drivers an enhanced driving experience, this includes an array of equipment that is increasingly in demand by drivers who are more reliant upon technology throughout everyday life (Walker et al., 2001). For example, there has been a recent shift towards wearable technologies, such as glasses and watches that link with smart phones, giving another information outlet. These devices may never have been intended for use in the vehicle, but their use by drivers needs to be considered. In line with Rasmussen's RMF (1997), legislation at the top level must be aware of these developments in the equipment (bottom of the framework), and the 'end users' expectations in order to adequately manage and regulate safety. This propagation through the system has failed in recent years, the technological advancement at the bottom levels is not yet widely accounted for in policies or design standards. Although research has been conducted into the potential effects of devices such as google glasses on driving performance (Sawyer et al., 2014), the disadvantage of distraction from these devices has yet to outweigh the potential benefits they could have in terms of monitoring the driver, providing corrective feedback and reducing load in multitasking (Sawyer et al., 2014; He et al., 2015). More thorough testing to understand the usability of all functioning of these devices is required before policy can be set. Yet, without a definition of distraction or proper control, evidence against the use of certain devices is hard to discern from research (Regan et al., 2011). It is evident that perhaps the effect other technologies may have in vehicles is not considered before their release. This means that legislation has to play catch-up, leaving road users potentially at risk until the safety implications are realised. This was an issue that was initially found with mobile phones (WHO, 2011).

As with hands-free devices, other technologies must abide by the same universal and national HMI guidelines, to ensure they are fit for use on the road (IC2 and R1). The design of the device and awareness of risks feeds down the system in a way similar to phone use. Yet, the drivers' risk perception is less well informed by media campaigns as they tend to focus only on the fatal four outlined by the Transport Select Committee (2016), of which mobile phone use is featured (alongside not wearing a seat belt, speeding and drink/drug driving), while other technologies are overlooked (C5). As all other devices are legal, they are likely to be thought of as less risky (e.g. Patel et al., 2008), in the same way as hands-free is thought to be of low risk as it too is permitted in vehicles (Young & Lenné, 2010) (EU5). Drivers tend to have a high willingness to engage with the devices and perceive them as assisting their driving performance and overall experience, in line with manufacturer's publicity. For example, satnav systems, whether built-in or an aftermarket device, are also now widely used by drivers who regard them to have a high 'point of need' (Burnett et al., 2004). Their assistance with the driving task increases their utility and reduces the perception that they may actually risk safe performance (Blomquist, 1986). As in Figure 5.3 with phone use, Figure 5.4 suggests that the perception of risk and the decision to engage with the device alters the adaption of driving performance and ultimately the maintenance of safe driving performance (EU5 and EU6).

The ambiguity in legislation on what does and does not conflict with the driving task is exploited with multiple devices being brought into the vehicle (i.e. built in and nomadic). Although the majority of these devices align with the HMI guidelines, the ambiguous information regarding the terms of use while driving mean that the driver is left unclear on the priorities they should make when engaging with the device. Manufacturers provide instruction manuals on best practise and guidelines on appropriate use to remove themselves from a position of liability, but often these are not read or are easily ignored by the end user (Mehlenbacher et al., 2002). This is an issue as the AcciMap shows how the driver is left widely responsible for the actions of the system. The high willingness to engage and the potential utility of technology is not balanced with the knowledge of their distractive effects (EU1). Often the utility and benefits associated with engagement overshadow the perceived safety impact (e.g. Kircher et al., 2011). The AcciMap can therefore be used to suggest that actors at the higher levels of the sociotechnical system are not doing enough to direct the end user on the goals that can conflict with the driving task and how to prioritise them. As the law directly relates to the use of the device in vehicles, it is the driver caught using them who is held responsible rather than those who permitted the devices to be brought into conflict with the driving task in the first instance.

5.3.3 Comparison between Mobile Phone Use AcciMap and Other Technology Use AcciMap

The AcciMap analysis allows legislation on phone use and other technology use while driving to be compared on their efficacy in managing distractions. Table 5.2 and Figure 5.5 highlight the number of distraction factors that are supported and unsupported within the AcciMaps (Figures 5.3 and 5.4).

TABLE 5.2
Frequency Table of Supported and Unsupported PARRC Factors across the Phone Use and Other Technology Use AcciMaps

PARRC Factors	Phone Use AcciMap Supported	Phone Use AcciMap Unsupported	Other Technology Use AcciMap Supported	Other Technology Use AcciMap Unsupported
Goal conflict	18	6	1	20
Goal priority	5	3	0	8
Resource constraints	4	1	4	2
Adapt to demands	3	1	3	1
Behavioural regulation	2	0	2	0
Total	32	11	10	31

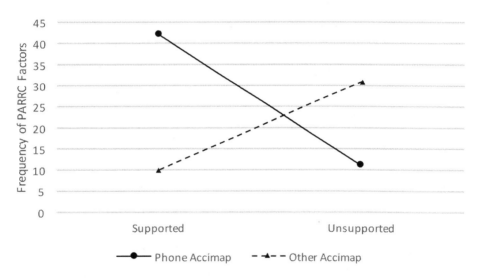

FIGURE 5.5 Graph showing the difference in the frequency of supported and unsupported PARRC factors in the phone use AcciMap and the other technology use AcciMap.

A chi-squared analysis (Table 5.3) revealed that the phone AcciMap and the other technology AcciMap were significantly different in the number of supported and unsupported PARRC factors, $\chi^2 (1) = 21.02$, $p < 0.01$.

Table 5.2 shows a notable contrast in the goal conflict factor between the two AcciMaps, with only one element found to support the system in mitigating distraction from goal conflict between the driving task and other technology. Furthermore, the need to prioritise goals to prevent distraction from other technology is not captured by any element in the other technology AcciMap. Where goal conflict and goal priority are supported, it suggests that legislation is giving clear direction on how devices must be managed when driving such that their use does

Legislation on In-Vehicle Technology

TABLE 5.3

Contingency Table Showing the Observed Totals, Expected Totals and χ^2 Statistics

	Supported			Unsupported			
	Total	Expected Total	χ^2	Total	Expected Total	χ^2	Marginal Row Totals
Phone	32	21.5	5.13	11	21.5	5.13	43
Other	10	20.5	5.38	31	20.5	5.38	41
Totals	42	42	10.51	42	42	10.51	84 (Grand Total)

not conflict with the primary driving task (Parnell et al., 2016). The specific ban on handheld devices distinctly implies the line between appropriate and inappropriate use of the device. Using the phone when driving will lead to the system migrating across the line of appropriate behaviour, inducing a risk to safety (Rasmussen, 1997). This line is less clear and more subjective within legislation regarding other technologies, for both drivers and law enforcers. It is often left to the driver to decide how to behave appropriately (e.g. Young & Salmon, 2015). Legislation can therefore be creating a false notion of safety by allowing some devices over others. Emphasising the risks of handheld phones and prohibiting them from vehicles may lead drivers to think that other devices are comparatively less risky if they are not given the same treatment. The ban on handheld phones may therefore be creating the conditions for distraction from other in-vehicle technologies to emerge from the system. Drivers often overestimate their performance and underestimate risks on the road (Wogalter & Mayhorn, 2005; Horrey et al., 2008), their ability to abide by these subjective laws is therefore called into question.

5.3.4 APPLICATION TO SPECIFIC EVENTS

Case study examples are used frequently within Human Factors research in order to relate theoretical models and psychological factors to the in-depth analysis of naturalistic behaviour. This has allowed further insights into behaviour within the aviation (Plant & Stanton, 2012), rail (Stanton & Walker, 2011) and road transport domains (Parnell et al., 2016). The AcciMap methodology is ordinarily applied to single events in order to assess the systems elements that contribute to the event outcome, which Hancock, Mouloua and Senders (2009) claim allows for deeper insights when studying human-machine interactions. To further highlight the differences between legislation on phone use and other devices, application of the AcciMaps to individual naturalistic scenarios were explored.

5.3.4.1 Scenario 1

A delivery driver's mobile phone rings while driving. The display on the phone informs them that it is their manager calling. This is acknowledged as an important call to take.

80   Driver Distraction

The safest option in this scenario is not to answer the phone. Indeed, according to the health and safety at work act, employers must inform anyone driving for work purposes not to use their phone while driving (Health and Safety at Work etc. Act, 1974). This should include targeting those making the phone call, knowing there is a likelihood that the person they are trying to contact is driving. Furthermore, employees should be made aware that hands-free devices also have the potential to distract, in line with the governments guidelines. However, as they are not banned from use in vehicles, they are seen as an alternative means through which to communicate with workers while driving. The driver in this scenario is highly motivated to answer; as more people drive for work, the need to remain productive while travelling is high, this is especially true of those in the sales industry (Eost et al., 1998). It is in the company's interest to increase productivity by allowing the driver to communicate when driving and therefore hands-free kits are useful to them as they do not require breaking the law. It is determined that, as long as warnings to employees are given, the responsibility is passed on. Figure 5.6 shows the employee is permitted to enhance their efficiency by using their phone in hands-free mode while driving; when they receive the call from their manager, they are therefore able to prioritise this call. As there may be consequences for not answering their manager's call, the driver is highly motivated to respond. The drivers' ability to adapt to the increased demand they are subjected to when engaging with the phone task will determine their ability to maintain safe driving performance. Yet, as they are permitted to use the hands-free device, they have a low perception of the risks involved. They are also motivated to engage with their manager, so cannot adapt the phone conversation and must adapt their driving behaviour to account for their reduced ability to perceive the environment (e.g. Alm & Nilsson, 1995). Here it can be said that the pressures placed on the employee to increase their efficiency while driving at work force the system into this position of unsafe practise. The system's behaviour pushes it closer to a point at which it is forced across the line of appropriate behaviour (Rasmussen, 1997).

Employers have a responsibility for employees' health and safety while at work and thus risks should be kept to a minimum. While it may go against the productivity targets of an organisation, safety of personnel should be paramount. To prevent actors at the end-user stage from being required to make decisions on engagement, avoiding the use of work mobiles in vehicles by employers, including hands-free use, would be the safest option. An example counter measure to this would be the use of applications that block out phone calls when it is detected the phone is in a moving vehicle, or "drive" settings built into the phone. Expectations of the employer and the pressures imposed on staff need to change.

5.3.4.2   Scenario 2

A driver is using a satnav to direct them to their destination when they are informed, by an intelligent road sign, that there is heavy traffic up ahead. They therefore wish to change their route to avoid the traffic, which means interacting with the satnav, using its touchscreen interface.

The satnav design meets all regulations set by higher level elements. It warns the driver before they enter the original destination at the start of their journey

Legislation on In-Vehicle Technology 81

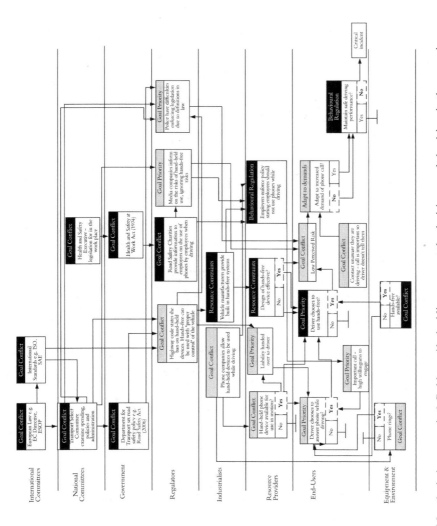

FIGURE 5.6 AcciMap depicting scenario 1. Bold type/dashed lines suggest the decisions made by driver in the scenario.

that the device is not permitted for use when driving, but it does not physically stop them from doing so, passing on the responsibility of any incident caused by engagement when driving to the driver. The touchscreen interface requires the driver to take their hands off the wheel and their eyes away from the road scene. Yet, it is highly useful in allowing the driver to get to their destination. Hence there is a high willingness to engage coupled with a low perception of risks, as shown in Figure 5.7. As the driver utilises the available technologies to prevent them suffering traffic delays, they progress the performance of the system towards an increased risk of incident. Taking the eyes and hands away from the driving task for more than 2 s greatly increases the risk of incident (e.g. Klauer et al., 2006). The warning upon start-up of the device is easily ignored when the motivation to interact with the satnav is high. The driver is effectively encouraged to test the boundaries to determine how they will perform when using the device while driving as the capacity of the driver to accurately perceive the appropriateness of their behaviour is impeded.

5.4 GENERAL DISCUSSION

This chapter has shown how applying the RMF (Rasmussen, 1997) to legislation regarding technology use by drivers with an AcciMap analysis can graphically represent the sociotechnical system and determine how distraction is an emergent behaviour of the system. The factors of distraction identified from the literature in Chapter 4 were applied as taxonomic references from which to categorise normal behaviour across multiple events. This overcame previously cited limitations of AcciMaps that they were not generalizable across multiple events, supporting recent research by Salmon, Goode, Taylor et al. (2017) and Trotter et al. (2014) in providing evidence to counter this limitation. This has enabled, for the first time, an investigation into the impact of UK laws on the distractive potential for in-vehicle technologies within a wider sociotechnical system perspective. Although the analysis in this chapter focuses primarily on UK legislation, many other countries consider controlling the use of handheld phones to be a priority in distraction legislation, with less specific guidance given to the range of other technologies not found in vehicles, meaning the findings are widely applicable.

Accident causation is now considered across many domains to be a consequence of the interrelationships within the sociotechnical system (Larsson et al., 2010; Leveson, 2011; Salmon, McClure & Stanton, 2012; Lansdown et al., 2015). The ability to infer causation therefore requires insight into the complexity of this system, a feature which accident causation models such as STAMP, HFACS and AcciMaps are able to tap into (Salmon, Stanton et al., 2011). Chapter 3 highlighted that the application of these methods to the road transport domain have been sparse. The cause and effect relationship between the driver and distraction is typically assumed in the assessment of road safety and the intervention methods that are applied. The use of cause-effect methods in studying driver distraction independently from the context that it occurs within has limited the ability to assess the impact of wider systems factors on the emergence of incidents resulting from driver distraction. The AcciMap analysis conducted in this

Legislation on In-Vehicle Technology

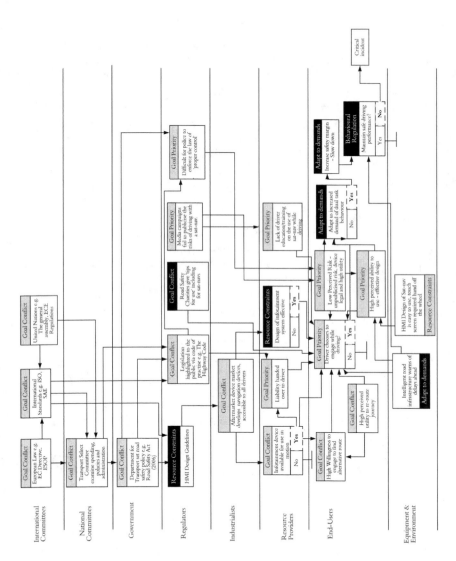

FIGURE 5.7 AcciMap depicting scenario 2. Bold type/dashed lines suggest the decisions made by driver in the scenario.

84   Driver Distraction

chapter has highlighted the benefits of systems-based methods and insights into the behaviour, without the direct observation of the behaviour in experimental conditions. This is, however, not to state that the use of experimental methods is irrelevant to assessing the role of systemic factors in driver distraction. Chapter 3 suggested that combining methods to enable insights into the progression of distraction-related events and the influence of systems factors should be applied in future research.

Utilisation of Rasmussen's RMF (1997) and the AcciMap approach has allowed legislation on in-vehicle technology use to be assessed from a systems approach, identifying the role of all relevant systemic elements. A complex combination of actors involved in the development of incidents caused by technological distractions have been highlighted, outside of those immediately involved in the accident, notoriously the driver (Tingvall & Haworth, 2000; Larsson et al., 2010). Furthermore, the location of these consequential elements within the sociotechnical system is highlighted which is of use to providing targeted distraction countermeasures (see Table 5.4). Taking a higher-level approach to assessing accident causation is important, as it allows the source of accident within the higher levels to be identified and targeted with higher level countermeasures for widespread change (Branford et al., 2008). This aims to move away from countermeasures that solely focus on targeting the driver. Here, the AcciMap methodology has been expanded upon to allow it to classify the actions of elements to failure factors, specific to the domain rather than generalised taxonomies which feature in other systemic accident causation methods, including STAMP and HFACS.

5.4.1 RECOMMENDATIONS

This chapter builds on the observations made in this book that legislation regarding driver distraction has focused heavily on the role of the driver, while ignoring the responsibility of the wider road transport system (e.g. Young & Salmon, 2015; Larsson et al., 2010). The outcome of this analysis has enabled the identification of possible recommendations which can be made to the sociotechnical system to target driver distraction from technological devices. These recommendations, stated in Table 5.4, go beyond the current methods applied to targeting drivers with increased consequences for engaging (e.g. Department for Transport, 2016a), or that indirectly point to the driver as the cause of error, to target the root cause of technological distractions.

5.4.2 EVALUATION AND FUTURE RESEARCH

A limitation of this AcciMap analysis was that it did not reference public opinion on technology use. Rasmussen's RMF (1997) states that government policies should be informed by members of the public, the end users, who must abide them. It is important that those who must abide by the laws accept them and those who regulate them. Analysis into opinions of drivers has not been as widely observed as quantitative measures of performance, yet Chapter 3 has identified that it is important to consider this, alongside other methodologies (Young & Lenné, 2010). Table 3.1

Legislation on In-Vehicle Technology

TABLE 5.4

Recommendations across Levels of the RMF Hierarchy

Hierarchical Level	Recommendations
International committees National committees Government	• Provide clearly worded legislation on avoiding driver distraction that relates to the whole system – not just the driver. This would require clearly defining driver distraction and making all elements aware of their involvement in its emergence. This includes device developers that provide a competing activity that does not enhance the drivers' safety. It also includes telecommunication providers, nomadic device designers as well as those creating the image that is provided in the media on the engagement with these devices, to highlight the message. • Clearly outline what 'proper control' of the vehicle refers to.
Regulators	• Ensure regulators set out clear guidelines to transport companies that state that they should have a safety management system set up that includes a systematic approach to distraction issues which relate to their operations. • Outline what resources may be constrained by devices and how this may implicate the driving task.
Industrialists	• Thorough testing of devices by drivers should be conducted before their widespread use to determine how they may influence attention towards the road. This should highlight all of the potential risks related to the use of the device. This can then facilitate exploration of systemic actions to minimise any uncovered risks. • Enforcement needs to effectively back up the legislation. Policing all drivers is not possible, monitoring of other system factors at an early stage is required such as device manufacturers, phone companies and employers. • Research is required into the driving, contextual and situational conditions drivers are able to engage with devices. This is likely to differ between devices depending on their resource demands and relevance to the driving task.
Resource providers	• Companies that contribute to the development of technology used within the vehicle, for example phone designer, navigation system designers, in-vehicle HMI developers, should not encourage temptation in the driver to use devices while driving. Legislation should place ownership of incidents on manufacturers who do not limit the use of distracting tasks while driving through freeze-out mechanisms, limited functioning or voice-activated alternatives when task cannot be performed safely while driving. • Preventing engagement with potentially distracting devices at inadequate times, as suggested through research in the level above, should be upheld by those implementing devices for use in the vehicle.
End-User	• The end users need to fully realise the risks of engaging with devices and how it may impact the driving task and the users' ability to perform the two tasks concurrently. The same level of risk should be attributed to distracted driving as drink-driving and speeding behaviour, making it a socially unacceptable behaviour. Other elements such as media, education and road safety charities can work to enforce this attitude change at the end-user stage.

(Continued)

TABLE 5.4 (*Continued*)
Recommendations across Levels of the RMF Hierarchy

Hierarchical Level	Recommendations
	• The high willingness to engage with devices needs to be offset by the potentially distracting effects of the device to make dangerous tasks less desirable to use when driving.
Equipment and environment	• The fast pace of technological change which is causing novel technologies to enter vehicles needs to be monitored. This means that the activities that are conflicting with the driving task are understood by other elements in the system such as the police who should look to enforce the law on new devices, media and road safety charities who should publicise the risks of the technologies as well and standards and guidelines which may need to be updated.
	• The conditions in the environment which impact the drivers' ability to engage with technological devices need to be recognised and the drivers' ability to interact with them at these points limited.

in Chapter 3 shows the different outputs that can be obtained from using different methodologies. This highlighted the role of subjective measures which aim to gain an understanding of the behaviour from the individual's viewpoint. Surveys have been used to ask drivers to rank activities in the vehicle on their perceived distractive effects and their perceived risk (Patel et al., 2008; Young & Lenné, 2010). Although, as these are quantitative subjective methods, they are limited by their inability to disclose open-ended responses from drivers on their true perspective. Yet, the use of focus groups by Lerner et al. (2008) sought individuals' motivations and risk perceptions in relation to mobile phone devices in the USA in 2002. They found that drivers did not attribute particular risk to engaging with phones while driving and were also more willing to use phones while driving than navigational devices, which was in contrast to survey-based methods. Yet, at the time the focus groups were conducted, there were no laws against the use of handheld mobile phones in the USA. An online survey conducted by Young and Lenné (2010) found drivers in Australia, where mobile phones were banned while driving, rated phone tasks to be among the riskiest activities to perform while driving. This could suggest the impact that legislation has on the drivers' risk perception and willingness to engage. However, the high ratings of risk drivers attributed to mobile phones in the online survey by Young and Lenné (2010) was not found to correlate to a reduction in their use while driving. Further research is required to understand the driver's perceptions of possible sources of distraction in the vehicle, the impact that legislation has on this, as well as the context that is created by the sociotechnical system surrounding the behaviour.

5.5 CONCLUSION

Rasmussen (1997, p. 184) stated *"society seeks to control safety through the legal system"*, yet this chapter has shown the inefficiency of legislation that does not regulate the appropriate actors across the whole system. In doing so, it leaves road users vulnerable and may inhibit the emergence of road safety. This chapter has highlighted the effect that current legislation regarding in-vehicle technology use while driving has on all of the actors across the road transport system that it relates to. Applying the PARRC factors to the AcciMap methodology has allowed a first step, within the road transport domain, in overcoming previous criticisms of AcciMap methodology. It has also highlighted the relevant actors which prevent the use of handheld phones while driving and the comparative lack of clarity in determining the other device drivers are permitted to engage with while driving. Recommendations across each level of the sociotechnical system have also been proposed, such as a need for more clearly worded legislation that can be easily enforced without subjectivity. These recommendations aim to prevent the responsibility for distraction to fall solely on the driver, motivating further work to determine solutions that apply across the whole system surrounding driver distraction from technological devices.

The need to review public opinion with respect to the driver distraction from in-vehicle technologies has been suggested as an avenue for future research, as this was not included in the development of the AcciMaps in this chapter. Methods are required that enable insights into the individuals' perspective on technology use in the vehicle and their interactions with other systems actors. While the focus in this chapter has been the top-down influence that legislation has on the sociotechnical system surrounding driver distraction, the bottom-up influence of the context surrounding the drivers' interaction with technology is the focus of the next chapter. This will focus on the drivers' perspective and the factors that they deem to influence the progression of distraction in relation to the other hierarchical levels of the system. This will utilise a semi-structured interview methodology with a sample of drivers that allows for open-ended responses to provide reasoning on why drivers are willing to engage with distractions.

6 Creating the Conditions for Driver Distraction
A Thematic Framework of Sociotechnical Factors

6.1 INTRODUCTION

Previous chapters in this book have sought to identify the systemic actors that influence the emergence of driver distraction related to in-vehicle technology with respect to the sociotechnical systems actors and their interrelationships. This has applied, and extended, the risk management framework (RMF) (Rasmussen, 1997) to explore the hierarchical levels of the system surrounding the behaviour (Chapter 5). The establishment of the Prioritise, Adapt, Resource, Regulate, Conflict (PARRC) factors of distraction in Chapter 4 and their application to an AcciMap analysis in the previous chapter has suggested how legislation may be creating the conditions for distraction. Yet, it was highlighted that the role of public opinion on the use of distractive technologies within the vehicle was not included. While the previous chapter highlighted the top-down influence that current legislation has on driver engagement with mobile phones, as well as other technologies, the bottom-up influence of the driver and the surrounding contextual factors that influence their interaction with technology requires further insight. Chapter 3 detailed that methods that focus on the cause and effect relationship between the driver and distraction fail to account for the wider context within which the behaviour occurs and the progression of distraction-related error in relation to the system. This chapter therefore seeks to obtain an understanding of the drivers' perspective on their engagement with technological devices while driving in order to understand how this may be influenced, or in turn influence, the interactions between systemic actors.

6.1.1 Voluntary Distraction: Theory and Methodology

An analysis of crash data conducted by Beanland et al. (2013) found 70% of distraction-related crashes to be voluntarily engaged by the driver. This highlights the importance of understanding the influence of the drivers' motivation and their intention to engage with distracting tasks. While legislation is imposed on drivers to prevent their interaction with tasks that may pose as distractions, drivers are still engaging with them. Previous research in the literature has informed the adverse consequences that technologies pose, yet the contextual and motivational factors that lead to engagement in potentially distracting tasks are

89

under-researched (Young & Regan, 2007; Young et al., 2008b; Young & Lenné, 2010; Horrey et al., 2017). Though it is useful to understand the effects of distraction on driving performance, the diversion of attention away from the safe activities of the driving task does not always result in adverse consequences (Stanton & Salmon, 2009). It is therefore important to understand the causal factors which lead to the diversion of attention, and how they may manifest in accidents (Young et al., 2008b; Young & Lenné, 2010; Lee, 2014).

An established and popular theory resolving to understand individual intention and motivation to perform behaviours is the Theory of Planned Behaviour (TPB; Ajzen, 1991). Surveys have been used to determine behavioural intention through targeting the key comprising factors deemed to influence intention: attitude towards the behaviour, social norms surrounding the behaviour and perceived behavioural control over the behaviour (Ajzen, 1991). Implementation of the TPB survey to drivers and their use of technology while driving suggests that the TPB factors account for 39%–43% of variance in intention to make a call on a mobile phone and 11%–13% of variation in text messaging, with attitude being the strongest predictor (Walsh et al., 2008). Conversely, Zhou et al. (2009) found that the TPB components accounted for 43%–48% variance in intention to use a hands-free device, which was most strongly predicted by perceived behavioural control. Thus, application of the theory to the behaviour of distracted driving has had mixed findings. Furthermore, a critique of the TPB survey is its lack of consideration for other factors, outside of the theory, that drivers themselves consider to influence their engagement in potentially distracting tasks.

Some research has been conducted into the decisions that drivers make to engage with distractions in simulators (Metz et al., 2011; Schömig & Metz, 2013), on test tracks (Horrey & Lesch, 2009) and through the analysis of data derived from naturalistic studies (Metz et al., 2015; Tivesten & Dozza, 2015). A challenge in the assessment of driver distraction research is the dichotomy between high levels of control and the naturalistic study of behaviour (Young et al., 2008; Burnett, 2009). While simulators offer control over external variables, such as road type and other road users, capturing realistic behaviour is compromised (Young et al., 2008). Yet, in naturalistic studies, the focus of data collection is on the driver and their triggered engagement with secondary tasks which allows them to collect data within the context that it occurs (Stutts et al., 2001; Neale et al., 2002; Klauer et al., 2006). Yet, this high level of validity is traded for a lower level of control (and thus measurement) of the contextual factors that influence drivers' engagement with secondary tasks (Metz et al., 2015).

To assess the factors that are impacting on the drivers' decision to engage with technologies, self-report methods can be used to capture the drivers' views on their own behaviour (West et al., 1993). Online surveys have been used to understand what distractions drivers engage with as they allow for large-scale data collection (e.g. McEvoy et al., 2006; Young & Lenné, 2010; Lansdown, 2012). Such studies have cited distractions sourced from technologies such as mobile phones, hands-free phones, satnavs and in-vehicle infotainment systems (IVIS) (e.g. Dingus et al., 2006; McEvoy et al., 2006; Young & Lenné, 2010; Harvey et al., 2011c). They have also provided insights into individual differences (e.g. McEvoy et al., 2006), the perceived risk of drivers when engaging in different tasks and their views on 'getting caught' (e.g. Young & Lenné, 2010).

Creating Conditions for Driver Distraction

The anonymity provided by accessing surveys remotely online may encourage honesty when asking questions that may require participants to reveal illegal behaviours that are characteristic of distraction-based research (e.g. using a mobile phone while driving, as discussed in the previous chapter). Yet, surveys are restrictive in their reliance on closed questions which can facilitate the imposition of the researcher's own agenda through their choice of survey questions (O'Cathain & Thomas, 2004). Closed-end questions, which have been favoured in the literature, limit the driver from detailing the influences they perceive to determine their decision to engage with technologies while driving. To more deeply understand *why* drivers become distracted, the application of more open-ended methods of qualitative data collection are required, the use of which have been limited in past research within the domain.

Huemer and Vollrath (2011) conducted short interviews with drivers at service stations which probed into their engagement with secondary tasks during their most recent drive. While face-to-face communication of this format allowed drivers to dictate their behaviour in an open manner, the large sample (289 drivers) meant that interviews only lasted 5 min and only sought to determine the prevalence of secondary task activity in the most recent trips taken by drivers. The only other research to the researchers' knowledge that has attempted to probe further into the drivers' choice to engage, using in-depth qualitative measures, was a focus group study by Lerner (2005). They conducted focus groups with drivers from four different age groups (teen 16–18 years, young 18–24 years, middle 25–59 years and older 60+ years) to discuss their willingness to engage with a variety of technologies that included a satnav, mobile phone and a personal digital assistant. They found drivers were primarily concerned with their motivation to perform the task (Lerner, 2005). Interestingly, it was found that drivers stated handheld mobile phone use to be safe to perform under most driving conditions and that they were motivated by social factors such as the use of their personal time. This is in contrast to more recent reports that have identified that drivers rate mobile phone tasks to be high risk and dangerous (e.g. Young & Lenné, 2010). This may be explained by the fact that Lerner (2005) conducted their focus groups in Washington, USA, in 2002, where handheld mobile phone use while driving was not yet prohibited by law. Furthermore, the use of focus groups may have facilitated social biases in what participants reveal with normative, cultural and dominance biases playing a role (Smithson, 2000). A more up-to-date, in-depth qualitative analysis is therefore required to understand the decision-making processes that drivers undergo when faced with modern technologies in the current sociotechnical climate and why they may, or may not, be motivated to engage with distractions.

The five key causal factors of the PARRC model ('goal priority', 'adapt to demand', 'resource constraints', 'behavioural regulation' and 'goal conflict'; Parnell et al., 2016), identified in Chapter 4, determine how in-vehicle technology may lead to distraction-related incidents relative to the interrelations between actors within the sociotechnical system. It is, however, unclear how the decision to engage with in-vehicle technologies while driving may be influenced by the causal factors of the PARRC model and wider systemic factors. While the application of the factors in the AcciMap analysis in Chapter 5 shows their utility in accounting for the systemic actors in driver distraction from technological devices, it still requires validation through the application to real-world behaviour. The causal factors that drivers deem to influence their decision to engage with distractions, and how this may result in distraction-related events, are of interest (Young et al., 2008; Young & Lenné, 2010; Lee, 2014). Furthermore, access to the systemic causal factors that influence the driver's engagement with in-vehicle technologies would enable the applicability of systemic countermeasures such as those provided in the recommendations section of Chapter 5 (Section 5.4.1).

6.1.2 OBJECTIVES

This chapter seeks to gain data from drivers on their self-reported reasons for engaging with technology while driving. The chapter is split into two studies. The first study sought to obtain the drivers' self-reported reasons for engaging with technology while driving using a semi-structured interview method to engage drivers in open-ended discussions on why they may be more, or less, likely to engage with various types of technology while driving. The use of inductive thematic analysis aimed to develop factors that drivers themselves deem to influence their engagement with technological tasks. The second study aimed to assess how the causal factors derived from the drivers in the interviews related to the causal factors that were developed from the literature in the PARRC model, which was generated via grounded theory in Chapter 4. The relevance of the drivers' reports during the interviews to the PARRC model of distraction allows the validity of the model to be assessed using data derived from drivers themselves. The findings also seek to assist in the provision of countermeasures that target the source of the issue, rather than observing with hindsight the effects of distraction.

6.2 STUDY 1

6.2.1 AIM

This study aimed to understand the drivers' self-reported reasons for engaging with technological devices while driving and the involvement of the sociotechnical system in their decision-making process. Previous research has sought to capture drivers' responses regarding their use of technologies using questionnaires and online surveys, yet this study aims to capture the drivers' subjective perspective in their own words. This will involve the use of semi-structured interviews to elicit discussions with drivers on their likelihood of engaging with different technological tasks across different road types.

6.2.2 METHOD

6.2.2.1 Participants

Drivers with experience on UK roads were specified, as the road types included within the semi-structured interviews related to those comprising the UK roadway system (Walker et al., 2013). A total of 30 participants were recruited (15 females,

Creating Conditions for Driver Distraction 93

15 males), across three age categories (18–30 years, 31–49 years, 50–65 years), with five females and five males in each category. Participants were required to hold a full UK driving licence with a minimum of 1 year of experience driving on UK roads (mean years of experience = 19.5, SD = 13.08). They were also required to be frequent drivers, driving on a regular, weekly basis in order for them to be exposed to situations where they may be inclined to engage with technology (mean hours spent driving a week = 9 h 45 min, SD = 6 h 20 min). Participation was voluntary. Further characteristics of the participants are given in Chapter 7 (Section 7.2.1.1).

6.2.2.2 Data Collection

To obtain the drivers' own views on why they engage with technological devices while driving, semi-structured interviews were conducted. Semi-structured interviews have been used effectively to investigate other aspects of driving behaviour (Gardner & Abraham, 2007; Simon & Corbett, 1996; Tonetto & Desmet, 2016), but they have not been applied to study how driver distraction is viewed by drivers. Their application within this research allowed for open-ended questions that enabled drivers to generate concepts they deemed important to their use of technological devices whilst facilitating a structured data collection method that could be reliably applied across all interviewees. They also allowed the researcher to probe into interesting concepts as they arose (Cohen & Crabtree, 2006).

The interviews were structured around a table that encouraged the driver to discuss their likelihood of engaging with a range of different technological tasks while driving across different UK road types in order to provide a discussion surrounding the situations and environments which may influence the use of different technological devices. Table 6.1 presents the list of technological tasks posed to participants in the interview. The tasks were drawn from the current literature investigating distraction from in-vehicle technology (e.g. Neale et al., 2005; McEvoy et al., 2006; Young & Lenné, 2010; Harvey et al., 2011c), as well as reports from road safety organisations and police reports (Department for Transport, 2015a; RAC, 2016). The tasks were chosen as ones that were frequently reported as distractions within this literature, without specific consideration for their duration or their complexity, as it was the decision factors relating to prominently identified distractions that were of interest. The road types presented have been shown in previous research to influence the drivers' situational awareness (Walker et al., 2013) and crash rates (Bayliss, 2009). They included motorways, major A/B roads, urban roads, rural roads, residential roads and junctions. Participants were presented with a road-type classification sheet with definitions, images and contextual information relating to each of the road types for clarity.

Within the table presented to participants, the UK road types identified by Walker et al. (2013) were placed along the top and the technological tasks were listed down the left-hand side. An extract from the table is shown in Figure 6.1. Participants were asked to fill in the boxes in Figure 6.1 by placing a number on a 5-point Likert scale relating to their likelihood of engaging in each of the technological tasks on each road type. One on the Likert scale represented 'extremely unlikely' and 5 'extremely likely'. While filling in the table, participants were also asked to verbalise their thought process and reasoning for their rating, including the factors that influenced it. While the Likert ratings allowed the driver to rate their likelihood, this chapter is not concerned with

94 Driver Distraction

TABLE 6.1

List of Technologies/Specific Tasks That Drivers Were Asked to Rate Their Likelihood of Engaging within the Interview Setting

Technology	Task
Navigation system	Monitor route
	Enter destination
Hands-free system	Find number from address book
	Answer a call
	Talk to other
In-vehicle system	Change climate control
	Change song/radio station
	Adjust volume
	Listen to music
	Verbally communicate with in-built system
Mobile phone/smartphone/portable device	Enter destination into navigation app
	Monitor navigation app
	Write/send a text
	Read a text
	Answer phone call
	Talk on the phone
	Enter/find a number
	Change song/audio track
	Use voice assist features
	Take a photo
	Use social media apps
	Check your email

Task	Road Type					Junction	
	Motorway	Major A/B road	Urban Road	Rural Road	Residential Road	Driving	Stopped
Navigation System							
Monitor route							
Enter Destination							
Hands-free phone							
Enter/find number							

FIGURE 6.1 Example extract of the table participants were asked to complete during the interview.

Creating Conditions for Driver Distraction

this data, but it will be the focus of the next chapter. This chapter focuses on the qualitative data generated from the participants' open-ended discussions surrounding their likelihood ratings. Participants were free to generate their own reasoning behind why they may or may not engage with a range of different technological tasks while driving across the road types. The researcher probed the participant to expand on their discussion points for clarity and further information where necessary. The same primary researcher conducted the interviews with all participants for consistency. Interviews were audio recorded, using an Olympus digital voice recorder, and transcribed.

A pilot study was conducted to determine if the technological tasks were representative of those used by drivers and to establish agreement on the descriptions of the UK road types. This revealed an overlap in some of the technological tasks, such as searching for a point of interest and a destination in a satnav. It also revealed that when drivers discussed their behaviour at a junction, they seemed to differ in their discussions surrounding the use of technology when stopped at the junction, for example at traffic lights, compared to when driving through an intersection. Therefore, the junction road type was split in two to represent both driving through a junction and stopping at a junction.

The interviews lasted approximately 30 min, although this varied depending on the discussions engaged by the participant and the researcher (average = 34.21 min, SD = 14.07). Interviews were audio recorded and transcribed. Due to the sensitive content, that is if they were likely to engage in activity that may be considered illegal under UK laws, such as using a mobile phone while driving, confidentiality was ensured to allow the participant to talk openly. The interview study was approved by the research institutes Ethical Research and Governance Organisation (ERGO reference: 24937).

6.2.3 Data Analysis

Transcriptions of the interviews provided the data set from which to analyse and draw inferences on the causal factors that drivers stated to have an influence on their decision to engage in the technological tasks while driving. Thematic analysis was used to organise, analyse and interpret key themes within the data (Braun & Clarke, 2006). A theme was defined as "...*something important about the data in relation to the research question, and represents some level of patterned response or meaning within the data set*" (Braun & Clarke, 2006; p. 10). The flexibility of the method is both an advantage, in facilitating adaptability across different approaches, and a limitation, due to comments of limited rigour from unclear methodologies (Braun & Clarke, 2006). Yet, Braun and Clarke (2006) comment that with clearly defined methods and commentary, thematic analysis can be a highly useful tool in drawing meaning from qualitative data. The methodology applied to the data set is therefore given in detail.

The thematic coding process was conducted in NVivo 11 software to add rigour to the qualitative research (Richards & Richards, 1991; Welsh, 2002). It also assisted in reviewing the thematic coding and sub-themes, as well as allowing for queries to be run on the coded data to interrogate the codes after they were developed.

6.2.3.1 Inductive Thematic Coding

The thematic analysis utilised a 'bottom-up' approach, whereby content was coded without a pre-existing framework, rather the framework developed through the analysis of the data (Boyatzis, 1998). Thus, the themes that developed were strongly linked to the source of the data (Patton, 1990). In contrast to deductive thematic analysis which seeks to look at aspects of the data that relate to the research framework under investigation, inductive analysis provides a richer insight into the data set as a whole, using naturally occurring themes (Braun & Clarke, 2006). It was not desirable to impose a framework on the driver's verbalisations, but rather draw on the concepts that the participants deemed to be important. Braun and Clarke (2006) state that the clarity of the methodology used to develop thematic code is essential to its validity which led to their own guidelines for thematic coding. Their guidelines were followed here, and the process of applying them to the data collected from the interviews is shown in Figure 6.2. The researcher conducted both the interviews and the thematic analysis, so that they were fully immersed in the data set and to enhance reliability. This does, however, mean that the researcher's own perspective must be acknowledged in the evaluation of this research.

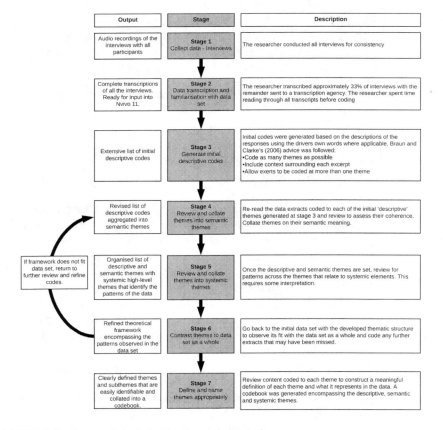

FIGURE 6.2 Stages of the inductive thematic analysis.

Creating Conditions for Driver Distraction

The iterative nature of inductive thematic analysis (see Figure 6.2) meant that the initial, descriptive, subthemes were coded as multiple individual concepts to draw as much information from the transcripts as possible (Stages 1–3, Figure 6.2). These were coded in the driver's own words, in vivo, to stay true to the data (Ritchie & Lewis, 2003) and to allow the emerging framework to reflect the real-life experiences and thought processes of drivers (Tonetto & Desmet, 2016). The multiple descriptive themes were then analysed, organised and refined into semantic themes (Stage 4). This required some interpretation and inductive theorising on the significance of concepts and their broader meaning (Patton, 1990).

The importance of identifying the systemic influencers of driver distraction was highlighted in the introduction, with current error taxonomies limited by ignoring the role of actors outside of the driver in error causation (Stanton & Salmon, 2009; Salmon et al., 2010). Therefore, in developing a causal framework of driver distraction, the role of elements in the road transport sociotechnical system were considered at Stage 5, where the relationship of the sematic subthemes to systemic actors was determined to develop a high-level systemic theme within the framework.

6.2.3.2 Inter-Rater and Intra-Rater Reliability Assessment

It was important to assess the reliability of the coding framework. Reliability relates to the consistency of observation, and it is important to the utility of the coding framework (Boyatzis, 1998). This can be achieved by asking other, independent, researchers to use the framework to code the interview transcripts (inter-rater reliability) as well as the primary researcher re-coding the transcripts using the framework after a passage of time (intra-rater reliability). Both of these methods were used here. 10% of the participants' transcripts were randomly chosen for rater reliability testing and given to the inter-raters and the researcher to code using NVivo 11. The original codes were removed so the raters could not see them. Two colleagues within the Human Factors research team were recruited to test inter-rater reliability. They were given the full thematic framework in a codebook (see Appendix A) and a 45 min training briefing where the interview study was explained and the themes were described before they were asked to independently code the same 10% of the transcripts. The raters were only required to code at the semantic level, while referencing the lower level descriptive themes to aid their coding. NVivo 11 software was used for the inter-rater coding with the researchers' coding hidden. Percentage agreement to the researchers' original coding was used to assess the reliability scores, other statistics could not be run as the data did not meet the assumption that all coding categories are mutually exclusive. Furthermore, where the aim of the research is more explorative, as the inductive analysis used in this research was, the use of percentage agreement is deemed acceptable (Kurasaki, 2000; Campbell et al., 2013). This method is widely applied in inter-rater reliability studies (Boyatzis, 1998; Plant & Stanton, 2013), and while there are still no established standards on the acceptable level of agreement between raters, Boyatzis (1998) deems 70% agreement as a necessary level of agreement. Inter-raters reached agreement percentages above this level (rater 1 = 81.24%, rater 2 = 74.87%).

Intra-rater reliability assessment was also conducted. The researcher re-coded 10% of the original transcripts 6 months after they had first coded the manuscripts within the initial coding process (detailed in Figure 6.2). This intra-rating assessment achieved a percentage agreement of 95.71%. Thus, indicating that both the intra-rater and the inter-raters were able to use the framework to code the data at a level that was much higher than chance, attesting to the reliability of the researchers' coding and the thematic framework.

6.2.4 Results

Inductive thematic analysis resulted in the development of a hierarchical framework of themes that reflected the drivers' self-reported likelihood of engaging in each of the technological tasks. An overview of the high-level systemic and semantic themes is shown in Table 6.2, the full thematic framework, including the descriptive, semantic, systemic thematic levels with example quotes, is presented in Appendix A.

It is evident from the full table presented in Appendix A that there was an extensive list of reasons that drivers gave for engaging, or not engaging, with the technological tasks while driving. A total of 168 descriptive themes were iteratively generated and revised into 18 semantic thematic categories. The generation of these themes was a lengthy but in-depth process. Clustering these semantic themes into higher level systemic actors that contribute to the occurrence of the causal factors gave another level

TABLE 6.2
The Systems and Semantic Subthemes of the Thematic Framework

Semantic Themes	Description	Example Quote
Driver	**References made by the driver including their mental/physiological state, experience, knowledge, skills, abilities and context-related behaviour.**	
(D1) Attitude of driver	Negative: Reference to negative attitudes of the driver towards performing the task while driving.	*Because, I just think it is the worst thing in the world, I just wouldn't do it, it's terrible*
	Positive: Reference to positive attitudes of the driver towards performing the task while driving.	*I don't see any problem with it personally whatsoever*
	Unnecessary: Reference to the driver perceiving the task to be unnecessary to perform while driving.	*That's something – it's just something that can wait until when you get home I think*
(D2) Tendency	Reference to the drivers' stated tendency to perform the task in the past and/or the future as an indicator of their likelihood to engage.	*I have been known to do that*
(D3) View of self	Reference to the drivers' stated view of themselves and their own behavioural tendencies when stating their likelihood to perform the task.	*If I am stopped I generally am a little bit more naughty*
(D4) Influence of others	Reference to other people and their influence on the driver and their likelihood of performing the task while driving.	*The shame if you did something bad, that everyone would think you are so stupid*

(Continued)

Creating Conditions for Driver Distraction 99

TABLE 6.2 (*Continued*)
The Systems and Semantic Subthemes of the Thematic Framework

Semantic Themes	Description	Example Quote
Infrastructure	**Reference to the specific road type within the road transport system, including the layout, contents, policy and regulated conditions.**	
(I1) Perceptions of surrounding environment	Reference to the context surrounding the road environment of a specific road type that is interpreted as an influencing factor in the likelihood to perform the task in the specific road environment.	*For these roads and junctions, it would require a lot more concentration*
(I2) Road layout	Reference to features of the specific fixed road environment that influence the drivers' likelihood to perform the task.	*Because to me a motorway, once you are on it, it is all moving in the same direction generally*
(I3) Illegality	Reference to the legislation on the use of the task while driving.	*I usually hold it in a low position, so police can't see*
(I4) Task-road relationship	Reference to the interaction between features of the road and the task that influence how the two may be compatible such that the likelihood of performing the task is influenced.	*Yeah it would be stilted, I would probably make the person on the phone aware, say hang on a minute but I would probably sound not as engaged*
(I5) Road-related behaviour	Reference to the actions and responses that are typical or required of the specific road type which influences the likelihood of performing the task on different roads.	*Urban road I think is more busy as well so I think the more sort of decisions you've got to make*
Task	**Reference to the details surrounding the specific task and engagement with it.**	
(T1) Complexity	Reference to the difficulty or ease of performing the task while driving.	*If you have to unlock the phone screen or whatever, etc., it is not as simple – well it is quite distracting*
(T2) Interaction	Reference to physical features of the task that relate to the interaction required to perform the task while driving. This relates to the interface design, device location and driver-required actions in order to engage with the task.	*It's only one button to press, so that's not an issue*
(T3) Duration	Reference to the time and/or length of the task.	*If it's a long text you might not read it*
(T4) Desirability	Reference to features of the task that influence how desirable it may be to perform while driving. This includes its use, performance or quality and alternatives available.	*I don't use my phone very much anyway so it's never been something that I have felt I have needed*
(T5) Engagement regulation	Reference to the factors that influence the conditions surrounding the onset of the task. They may relate to the physicalities of the task and/or the drivers' motivation relating to the task.	*I will always figure out what I'm going to listen to and set it going before I leave*

(*Continued*)

100 Driver Distraction

TABLE 6.2 (*Continued*)

The Systems and Semantic Subthemes of the Thematic Framework

Semantic Themes	Description	Example Quote
(T6) Ability to complete	Reference to features of the task which influence its ability to be completed in full while driving.	*because I've had the car for ages, I know where the switches are*
Context	**Reference to the circumstances surrounding the behaviour described.**	
(C1) Journey context	Reference to circumstances that form the setting for a journey that may influence the likelihood· to perform the task.	*if I am in a strange city, I would be less likely to mess around because I don't know where I am going*
(C2) Task context	Reference to circumstances that form the setting for the use of the task that influence the drivers' likelihood to engage with it.	*It's stuff when I actually feel like I need to send a message quickly, so if I've agreed to come home at a certain time and I'm running late for instance*
(C3) Road context	Reference to circumstances that form the setting surrounding the road in general (not related to specific infrastructure) that influence the likelihood to perform the task.	*I think it would be situational dependent, just how busy is it? I think*

to the framework that readily demonstrates the contribution of the system within which driver distraction occurs. In line with previous individual focused approaches, the driver emerged as a key actor. The driver category suggests that the driver is influenced by their own attitudes (D1), perceptions of themselves (D3) and their own tendencies (D2) in their engagement with distractions while driving, as well as how they feel they may be viewed by others (D4). Yet, the development of the other categories suggests that they are also influenced by other systemic actors.

The role of infrastructure was reported when responding to the different road types that were posed to the driver during the semi-structured interviews. Their perceptions of the road environment (I1) altered how likely they would be to engage due to the requirements of the driving task in these conditions, for example increased concentration required at junctions or reduced perception of risk on motorways. Road layout across road types (I2) was also widely discussed with the discussion of corners, road turnings and road visibility stated as elements contributing to the decision to engage. Drivers also made connections between the road and their behaviour on it (I5), as well as between the task and its use in relation to the road (I4). For example, the speed of particular roads or the availability of places to stop was discussed as contributing to their decision to engage while driving. Furthermore, the behaviour required in the driving task was also reported to influence their ability to engage in the different secondary tasks while driving, for example driving on a motorway was deemed to be easier by some drivers which motivated their engagement in more complex secondary tasks, than if they were in a more complex road environment. Illegality (I3) was a theme that is also mentioned

Creating Conditions for Driver Distraction

by drivers and is included in the road infrastructure theme, as it is in Stanton and Salmon (2009). Interestingly, the law was only one of the 168 other factors that drivers stated influenced their decision to engage. This highlights the potential for the development of other techniques to tackle the numerous other contributing factors, as a fear of the law was only a small contribution to the drivers' self-reported causal factors.

The task itself generated discussion on how it influenced the likelihood of the driver to engage with it while driving. Notably, this included how long the task would take to complete (T3), the method through which they could interact with it (T2), the complexity of this interaction (T1), its desirability (T4), the ability for the task to be completed (T6) as well as how its onset may be regulated (T5). The task theme has the most semantic subthemes, highlighting the number of variables relating to the task that influence the drivers' engagement with it. It was evident that drivers were aware of the differing complexities and ways of interacting with the different technological tasks and how they could manage these while driving. The role of the manufacturer in facilitating engagement and the influence of developments in HMI design concepts was particularly evident throughout these discussions.

The other systemic category was the wider context within which drivers discussed their engagement with the technological tasks. They discussed the type of journey (C1) that they may be going on that may require them to engage with the task more. For example, the use of a satnav was more likely on a longer journey when they didn't know where they were going or a phone call may be more likely if they were commuting. The context of the task itself (C2) was also discussed with the importance of the task to their priorities referenced frequently. Notably when discussing phone-based tasks, drivers reported that it was who they were communicating with, and their perceived importance of the communication, that greatly influenced their engagement with the task.

6.2.5 Discussion

The development of the thematic framework in Table 6.2 is the first attempt to develop an extensive list of the drivers' self-reported reasons for engaging with technological tasks. Previous efforts to assess the causal factors of driver error have suggested that the key factors relate to the following systemic elements: road infrastructure, the vehicle, the driver, other road users and environmental conditions (Stanton & Salmon, 2009; Salmon et al., 2010). Thus far, the development of error taxonomies has been heavily theoretical, emerging from the aggregation of previous literature and accident reports (e.g. Stanton & Salmon, 2009). The hierarchical levels of the framework that were inductively generated gave an insight into the higher level factors that are closely tied to the drivers' own comments and discussions. The high-level factors suggest the importance of the driver, the road infrastructure, the task and the wider context on the drivers' decision to use the technological tasks while driving. Differences to the causal factors taxonomy presented by Stanton and Salmon (2009) include a more specific focus on the task in the thematic framework rather than the vehicle as a whole. This is likely due to the design and aims of the study which required drivers to talk through their likelihood of engaging with a variety of different technological tasks. Had the participants been asked to drive or talk more about the interaction between completing the task while driving, more references to the vehicle may have emerged (Pedic & Ezrakhovich, 1999). 'Other road users' is also absent from the high-level themes of the thematic framework, but does appear within the infrastructure theme under the 'perceptions of surrounding environment' semantic subtheme (see Appendix A). This could suggest that the drivers' views of other road users are tightly linked to infrastructure and the surrounding environment when deciding to engage with technological tasks. Yet, other research conducted in a naturalistic driving study has suggested that other vehicles in front of the driver do not influence drivers' decisions to engage with technological tasks (Tivesten & Dozza, 2015). Further research is therefore required to determine the role of other road users (See Chapter 8).

While there was an evident involvement of systemic actors that influenced engagement, the driver emerged as a key systemic theme due to the references that participants made to their attitudes, perceptions and views on the technological tasks which influenced their likelihood of engaging. This compliments other research that utilised surveys to identify that drivers' intention to engage is influenced strongly by their attitude towards the behaviour and their perceived risk of the task to driving performance (Walsh et al., 2008; Zhou et al., 2012). The voluntary aspect of distraction (Beanland et al., 2013; Lee, 2014) and its self-regulatory association (Tivesten & Dozza, 2015) are inherent to the behaviour, yet this should not be studied independently to the wider context and system within which the behaviour occurs (Young & Salmon, 2015; Parnell et al., 2016).

The road infrastructure was discussed extensively, leading to multiple themes within the framework. This compliments the research conducted in a naturalistic study conducted by Tivesten and Dozza (2015) who found the drivers' ability to anticipate the road infrastructure, such as tight corners or straight roads, influences the drivers' engagement with distracting tasks. While they suggested that other road conditions did not influence the drivers' engagement as they could not be anticipated, the findings from this interview study suggest that road environment and the relationship between the task and the road is discussed as a causal factor in engagement and thus drivers can anticipate the effect it may have on their driving performance, within the interview setting. Other themes in the framework have also been suggested in the literature such as task context (Lerner et al., 2008), task capabilities (Zhou et al., 2009) and journey context (Tivesten & Dozza, 2015). Yet, the aggregation of factors which have been inductively generated from a sample of drivers is novel and has strong theoretical applications.

The extensive range of causal factors within the framework includes the contribution of legislation to the drivers' decision to engage. Yet, there are a host of other contributing factors that suggest the potential for other measures through which to tackle the drivers' engagement with technologies. The PARRC model of distraction (Parnell et al., 2016; Chapter 4) highlights the relevance of systemic actors to the causal factors that are attributed to driver distraction that were generated from the literature. The development of the thematic framework in Study 1 offers the possibility to contrast the causal factors that drivers report in the interview study to those that are reported in the literature. This is explored within Study 2.

6.3 STUDY 2

6.3.1 Aim

The inductive thematic analysis conducted in Study 1 allowed the causal factors that influence the drivers' likelihood of engaging with technologies to be directly linked to the drivers' discussions. The PARRC model of distraction sought causal factors directly from the literature using grounded theory methodology (Chapter 4; Parnell et al., 2016). Yet, like the thematic framework developed in Study 1, it highlights the involvement of the wider sociotechnical system in the development of distraction. The aim of Study 2 was to determine the relation of the drivers' reports to the claims made in the literature by applying the thematic framework in Table 6.2 to the PARRC model of distraction developed in Chapter 4. This will seek to assess the validity of the PARRC model through triangulation with its application to alternative data sources (Hignett, 2005). It may also determine areas of the drivers' reports that have not been studied in the literature and therefore seeks to promote future research, as well as providing sociotechnical systems recommendations to the mitigation of distraction.

6.3.2 Method

The PARRC model of distraction was reviewed to assess how the literature-driven mechanisms relate to causal factors stated by drivers. The process through which this was achieved is detailed in Figure 6.3.

The relationship of the semantic causal factors stated by drivers to the key factors identified from the literature in the PARRC model was reviewed through discussions with two subject matter experts with over 40 years of Human Factors experience (Stage 1, Figure 6.3). The relationship of the semantic factors to systemic actors (Stage 5 of the framework in Figure 6.2) meant that the contribution of the systemic actors to the causal mechanisms of the PARRC model could be identified.

Interconnections are important within systems models, as sociotechnical systems emphasise the emergence of safety from the complex interactions between systemic elements (Leveson, 2004). The interconnections in the original PARRC model were derived from empirically tested connections made in the literature through associations made by authors in relating concepts to one another. Connections between the causal factors reported by the drivers in the interview study in Study 1 were identified using a matrix query in the NVivo 11 software that was used to code the data. Matrix queries allow the number of co-occurring coded themes to be quantified and highlighted, to determine the number and type of data excerpts that relate to co-occurring themes of interest. Stage 3 of the inductive thematic process (Figure 6.2) states the process for generating the initial descriptive codes within the data. This process requires excerpts to be coded to multiple themes (Braun & Clarke, 2006), allowing co-occurring themes to be reviewed afterwards. The linking of the semantic subthemes to the PARRC model mechanisms in Figure 6.3 allowed the links between the subthemes representing each of the PARRC factors to be explored. The total number of interconnecting statements in the interviews between the PARRC mechanism subthemes was calculated. The connections could then be reviewed through the NVivo 11 software to determine what the drivers' statements included that were

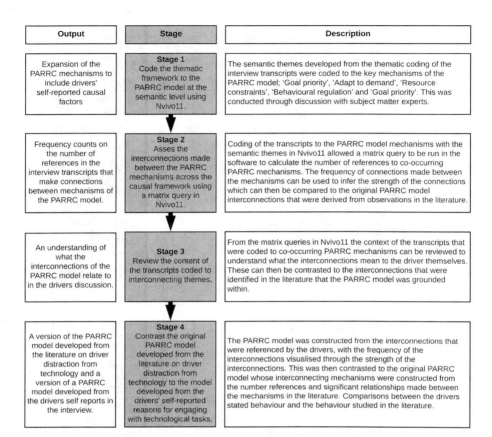

FIGURE 6.3 Stages in the application of the thematic framework to the PARRC model of distraction.

coded to the interconnecting PARRC mechanisms (Stage 3, Figure 6.3). Comparisons could then be made between the original PARRC model, grounded in the literature, and the reconstructed PARRC model built from the drivers' self-reported discussions on their reasons for engaging with technology while driving.

6.3.3 Results

Application of the semantic themes detailed in Table 6.2 to the PARRC factors and assessment of the interconnections referenced by the drivers led to the construction of the PARRC model developed from the drivers' self-reported reasons for engaging with distractive technologies while driving (Figure 6.3, Stage 4). This is presented in Figure 6.4. Insights that were gained from the application of the PARRC model framework to the themes identified and inductively generated from the interviews are discussed.

Each of the PARRC factors were found to be represented in the thematic framework developed from the drivers' self-reported reasons for engaging with different technological tasks while driving. The relevance of these themes to the factors is discussed below.

Creating Conditions for Driver Distraction 105

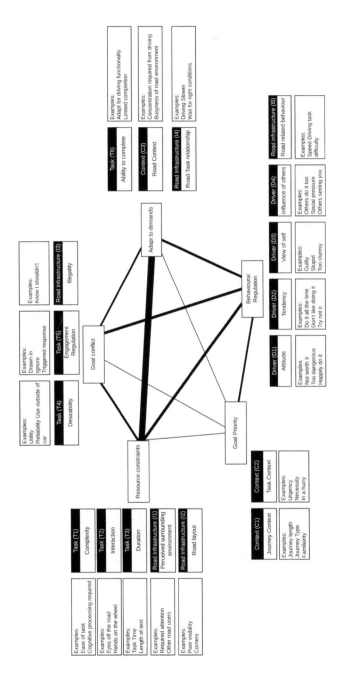

FIGURE 6.4 Application of the thematic framework, referenced in Table 6.2, to the PARRC model of distraction.

6.3.3.1 Adapt to Demands

Definition: The increased mental and physical demand associated with engaging in secondary tasks while driving requires adaption of either the primary or secondary task, or both for effective performance.

Example Quote: I think it's only when I've got to think a lot and I don't know where I'm going, that I would say to someone, 'Hang on, I need to concentrate on the roundabout, I need to concentrate on something' and I'd close the call (Participant 16, Talking on hands-free phone).

Drivers supported the notion that they adapt both their behaviour in the driving task and the secondary task in line with increased mental and physical demand when discussing their likelihood of engaging with a technological task while driving. The semantic subtheme 'Road context' (C3) highlights the need to alter and adapt their driving behaviour in line with the changing demands of the road environment. The semantic subtheme 'Road-task relationship' (I4) suggests that drivers also adapt their behaviour across different road types, such that they are aware of the different demands of different roads and alter their engagement with technology accordingly. In terms of the adaption of the technological task, the subtheme 'ability to complete' (T6) suggests that drivers adapt the functionality of the task, adjusting the completeness of it as illustrated in the example quote, in order to meet the demands of the driving task.

6.3.3.2 Behavioural Regulation

Definition: The self-management of attention, effort, attitudes and emotions to facilitate goal attainment.

Example Quote: You're not actively doing much when you're on the motorway. Same with the major A and B roads, so there's not a lot of actual steering it's just about focusing on what's ahead of you. So, if you're right behind someone then that's not a good thing to do but otherwise, yes (Participant 6, Answering a hands-free call).

The interviews provided information on the cognitive thought processes of the driver and their perceptions of their own behaviour relative to the surrounding environment. It has shown that the drivers' 'attitude'(D1), 'tendencies' (D2), 'view of self' (D3) and the 'influence of others' (D4) are key factors that relate to the regulation of their behaviour with respect to engaging in technology whilst driving. Furthermore, as is demonstrated in the example quote (Table 6.2, I4), the behaviour of the driver is also shown to be regulated by the road infrastructure within the task-road relationship theme (I4), with drivers discussing how they regulate their behaviour in relation to the road environment and attainment of the driving goal, which is altered across road types.

6.3.3.3 Goal Conflict

Definition: The existence of two or more goals that come into competition with each other, such that both cannot be completed concurrently without disrupting one another.

Example Quote: Use voice assistant features. Yes, I did it because my car was... it was because it was basically without petrol, so I needed to find a petrol station as soon as I could, so I asked Siri to find me the closest petrol station, and I've done that while I was driving. (Participant 15, Voice assistant features on mobile phone).

Creating Conditions for Driver Distraction

Drivers discussed the limitations of responding to co-occurring driving and technological task goals with respect to the features of the technological tasks (T4 and T5) and their knowledge on the laws (I3) which states that the conflicting goals should not be achieved in unison (i.e. driving while using a handheld phone under UK law). Features of the technological task which related to its potential to conflict with the driving task were its 'desirability' (T4) and 'engagement regulation' (T5). Technologies within the vehicle have developed over time to provide novel interactions and functionalities to the driver that were not previously available. This makes them desirable to would-be users (Walker et al., 2001) and therefore places them in conflict for attention with the main driving task. Drivers discuss the utility (T4) of the technologies and how this relates to their use while driving, as illustrated in the example quote (Table 6.2, T4). They also discuss the features of the task that regulate how able they are to regulate the onset of the task (T5) and the arising conflict this may have with the driving task. For example, many drivers commented on the triggered response that occurred when they receive a text while driving, for example *"Reading a text', you see I would read a text just because of the nature of the fact that it flashes up on your phone"* (Participant: 7, Task: Read text on mobile phone). This suggests that drivers did not always wish to engage in the technology task but the design of the device allowed it to compete for the drivers' attention, diverting it away from the driving task and providing a conflict for attention.

6.3.3.4 Goal Priority

Definition: The multiple goals that drivers face cannot be completed simultaneously; they must be prioritised in accordance with goal hierarchy. It is important that the priorities match the current demands to maintain safety.

Example Quote: Read a text on a hand-held device? Possibly, if you're waiting for something, maybe a – waiting for someone to give you a nod to go somewhere (Participant 7, Read text on mobile phone).

Prioritisation was found to be influenced by the context and circumstances which surround the interaction, not just the road infrastructure but the specific circumstances that may, or may not, lead to interaction with technology. The journey type and/or length (C1) influenced the requirement to engage. For example, longer journeys may lead to more interactions with the music system. The familiarity (C1) with the route was also suggested to alter confidence in prioritising the technological task. The circumstances surrounding the technological task were also important (C2), such as how important or urgent the task was, as highlighted in the example quote. The contextual factors suggest there are many situations in which engagement with technology is more or less likely to occur. These do not relate to road type or environment but specific moments and circumstances that cannot be foreseen and directly impact on driver's willingness to engage.

6.3.3.5 Resource Constraints

Definition: Attentional resources are finite; successful driver behaviour involves manipulating resources to enable their efficient distribution between tasks and according to the situational demands.

Example Quote: It pulls it up basically, next to the speedometer what song you are on, so that is the sort of looking away that isn't particularly distracting, because you look at your speedometer as part of normal driving don't you? (Participant: 10, Task: change song/radio station on IVIS).

The attentional resources of the driver were reported to be constrained by features of the task (T1–T3) and features of the road environment (I1–I2). The road types presented to the drivers led them into a discussion on the features of the road that influences their decision to engage with technology. Discussions on the road layout (I2) and their interpretation of the surrounding environment (I1) highlighted the drivers' awareness of the elements of the road environment which may limit their attention. Likewise, there was also an in-depth discussion on the characteristics of the technological tasks presented to drivers and how features such as its 'complexity' (T1), method of 'interaction' (T2) and 'duration' (T3) influenced the attention that it required. In some cases, such as the example quote above, the perceived resources required to interact are minimal which increases the likelihood of engaging, whereas for other tasks the perceived resources are too great to complete the task while driving.

6.3.3.6 Interconnections

Just as the factors of the PARRC model were founded in grounded theory methodology, the interconnections between them were identified from the empirically tested connections made in the literature (Chapter 4; Parnell et al., 2016). The interconnections of the PARRC model therefore require validation to determine their efficiency in capturing real-world behaviour. The interview transcripts provide details of the drivers' own views on their likelihood to engage in the technologies while driving and their decision-making processes in relation to a variety of technological tasks across all UK road types. They therefore form a data set from which to compare to the original empirically determined connections.

The linking of the semantic subthemes to the PARRC model factors in Figure 6.4 allowed the links between the subthemes representing each of the PARRC factors to be explored. The total number of interconnecting statements in the interviews between the PARRC mechanism subthemes is shown in Figure 6.5b. This is contrasted to the interconnections found from the grounded theory approach in the original development of the PARRC model in Chapter 4 (Figure 6.5a). The numbers on the interconnection in Figure 6.5a relate to the number of studies empirically testing the relationship between the factors. The size of the connecting lines represents the number of connections made in both diagrams. The matrix coding of the driver interview data only states the frequency of co-occurring subthemes, not the direction of any relationship that may occur, so the connections are shown as lines rather than directional arrows in Figure 6.5b.

Figure 6.5 contrasts the interconnections between the PARRC factors in the original model (Figure 6.5a) and the one based on the drivers' coded transcripts (Figure 6.5b). Aside from the quantities of total interconnections being higher in the interview-driven connections (Figure 6.5b), which is likely to be due to rich data source of the interviews in contrast to the 33 studies identified in the literature

Creating Conditions for Driver Distraction

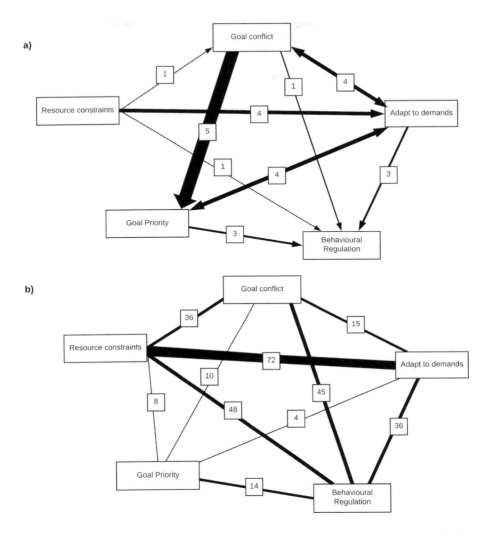

FIGURE 6.5 PARRC models developed from (a) empirically tested relationships in the literature and (b) the interview transcripts.

review (Chapter 4; Parnell et al., 2016), there are similarities in the model configurations. This suggests support for the underlying literature that the PARRC model is grounded in as the research is targeting concepts that are also generated to be important by drivers.

To assess the interconnections in further detail, the transcripts that were coded to the connecting themes in the matrix query are presented to understand how the drivers' reports may differ to those studied in the literature (e.g. Stage 3; Figure 6.3). These are shown in Table 6.3.

TABLE 6.3
Interconnections between the PARRC Factors as Drawn from the Interview Data

Connection		Description	Quote
Resource constraints	Goal conflict	Constrained resources mean drivers cannot perform two tasks at once so they are in conflict with each other.	*Well, again it is just I would find it too distracting, I wouldn't be able to drive and operate the software on my phone to enter the destination, because I wouldn't be able to see properly!*
Resource constraints	Adapt to demands	Limited resources require adaptation of behaviour.	*So the picture that you've got has lots of cars on the side of the road and houses and I'd be thinking, "Ah a car is going to pull out in front of me" or "I'm going to get very close to a car", so therefore 100% needs to be on the road at that point*
Resource constraints	Behavioural regulation	Drivers must regulate behaviour in line with their limited resources.	*Yes, I don't think that can be done in a safe amount of time. It would have to be less than a second I think and even then, even that on a motorway is risky. But I don't think I could do that in less than about 20 seconds, even if I knew the postcode and everything. So I just wouldn't*
Goal conflict	Goal priority	Drivers prioritise to overcome conflict.	*So also depending on who is calling, so if for example work is calling I'll probably turn that off or some number that I don't know, whereas if it's someone that I'm more likely to know or someone who's more likely to tell me something important then I will answer more readily*
Goal conflict	Adapt to demands	Drivers can adapt to the demands of the conflicting tasks.	*I'd be highly likely to do this on a motorway, because I think it's quite easy, again like using a phone I would do it staggered touch something look up, touch something else, look up*
Goal conflict	Behavioural regulation	The ability to regulate behaviour towards new tasks that conflicts the driving task influences the conflict for attention between the tasks.	*Just because it's more of a – for me, for some reason it's an automatic response, so the phone flashes and you naturally just look over and then you read the text*

(Continued)

Creating Conditions for Driver Distraction

TABLE 6.3 (*Continued*)

Interconnections between the PARRC Factors as Drawn from the Interview Data

Connection		Description	Quote
Adapt to demands	Goal conflict	Drivers adapt behaviour which can lead to further conflict between goals.	*Yeah, into the car but I listen to play lists so it's never really... actually, on the motorway it's finding a song, if I'm finding a song, searching on Spotify whereas if I was on a rural road or a residential area, I'd just be skipping a song*
Adapt to demands	Goal priority	Drivers can adapt to demands by prioritising their goals.	*The only time I might do it is at a junction, stopped; there's a chance I might if I really need to make a call, but otherwise I wouldn't moving, no*
Adapt to demands	Behavioural regulation	Drivers can regulate their behaviour to adapt to the demands of the environment and the task.	*yeah, ok so changing climate controls I definitely would on the motorway, as long as you are just cruising along*
Goal priority	Adapt to demands	The prioritisation of goals requires adaption of the primary and secondary tasks in line with current demands.	*Rural road probably, it depends how rural, if it's a really tiny road I probably don't answer it because it is quite nice to be able to hear the road*
Additional connections			
Behavioural regulation	Goal conflict	The drivers' attitude towards the task and stated intention to engage with it while driving influence the potential for the technological task to come into conflict with the driving task.	*No, never, never, never, never, never. Read a test? Honestly, I would never do this stuff*
Resource constraints	Goal priority	Drivers lend resources to the driving task to determine its priority before deciding to engage with it.	*I would never open up a whole message, but I might glance down and look at who it is from at least and what is written on it*

The interconnections provided validation for many of the interconnections that were included within the original mechanisms of the PARRC model that was generated from the literature in Chapter 4. Yet, some differences can be seen between the two models in Figure 6.5. These were assessed by looking towards

112   Driver Distraction

the transcripts that were coded to the connecting themes in the matrix query. They
include the following:

- *A reduction in the prominence of the connection between 'goal conflict'
  and 'goal priority'.*
  In the development of the original PARRC model connections, the literature was found to frequently consider the need to prioritise the goals of
  the driving task and the secondary task in order to resolve any goal conflict
  that may be occurring between the two (e.g. Dogan et al., 2011). Yet, the
  transcripts of the drivers' verbalised thought process suggest prioritisation
  of goals to be less of a concern when contrasted to other connections.
- *An increased prominence of the connection between 'resource constraints'
  and 'adapt to demands'.*
  Adaption of behaviour to manage resources was a well-represented
  notion in the interview transcripts, with 72 statements connecting resource
  constraints themes to adapt to demand themes. The connection was originally deemed to reflect the idea that adaption relates to the attentional
  resources that are available (Chapter 4; Parnell et al., 2016). The PARRC
  model aligns with the theory that the attentional resources of drivers are
  limited, such that both the primary and secondary tasks must compete
  for available resources to facilitate their effective performance (Wickens,
  2002). One way of ensuring that attentional resources are efficiently allo-
  cated across the primary and secondary task is through adaption (Noy,
  1989; Cnossen et al., 2000). Figure 6.5b and the example quote given in
  Table 6.3 suggest that drivers do report the adaption of their behaviour in
  order to facilitate their secondary task goal. It also suggests that this adaption to facilitate multiple goals is more important to drivers than the need to
  prioritise one goal over another.
- *The addition of a connection between 'resource constraints' and 'goal priority', which was not found in the literature that the original PARRC model
  was developed from.*
  Another key difference between Figure 6.5a and 6.5b is the addition of
  an extra connection between resource constraints and goal priority, which
  was not found in the literature that the PARRC model was drawn from and
  is therefore absent in the original model (Figure 6.5a). In Chapter 4, it was
  reasoned that the absence of this connection could stem from a number of
  causes which relate to the other connections in the model. These include
  the prominent connection between 'goal priority', 'adapt to demands' and
  'resource constraints', which prevents resources from constraining the prioritised goal, and facilitates adaption of behaviour in line with resource
  availability in order to prioritise one goal over another. Yet, the findings
  from the interview data suggest that drivers do make a connection between
  the resource constraints factor and goal priority, although it is minimal.
  Table 6.3 gives an example statement of the connection made by the drivers and suggests the connection relates to drivers lending some attentional
  resources towards the task in order to determine if it requires prioritising.

Creating Conditions for Driver Distraction

This is of interest as it suggests a concept that was not previously considered in the literature from which the original PARRC model was conceived.

- *An increased prominence of connection between 'goal conflict' and 'behavioural regulation'.*

Figure 6.5a suggested the connection between 'behavioural regulation' and 'goal conflict' reflected the bottom-up process through which certain goals may result in triggered responses that realigns the conflicting goals (as suggested in Chapter 4). The interview data presents a stronger connection between 'goal conflict' and 'behavioural regulation' in Figure 6.5b. A review of the coded transcripts connecting the mechanisms suggests it may reflect a connection in the opposite direction to the original PARRC model, going from 'behavioural regulation' to 'goal conflict'. The example quote from Table 6.3 highlights the influence that drivers' attitudes towards the technological tasks can have on their potential to conflict with the driving task. The driver in the example is adamant that they will never use the device while driving, stating they will turn it off so that it will not pose a conflict to their driving goal. This illustrates a reoccurring theme within the interview transcripts, many drivers held strong attitudes towards the use of technological tasks while driving, stating that they would never ever attempt to engage in the task while driving.

6.3.4 DISCUSSION

Study 2 has presented the application of the thematic framework, developed within the interview study in Study 1, to the PARRC model of distraction which presents the causal factors involved in driver distraction as stated in the literature. This has validated and extended the original PARRC model and the literature within which it is comprised. It has shown how the drivers' reports relate to the study of the behaviour in the literature. Driver distraction can be a difficult behaviour to study in its natural environment due to other confounding factors and the ethical issues of exposing participants to distracting activities (Carsten et al., 2013). Therefore, capturing the drivers' reported behaviour through open-ended discussions and establishing its relation to the existing literature provides the much needed validation of the research.

The causal factors reported within the interviews and their association with the PARRC model factors supports other research from naturalistic driving studies. The notion of driver adaption is particularly evident with studies that have looked at drivers' engagement with secondary tasks, with drivers slowing down (Metz et al., 2015) and increasing headway (Tivesten & Dozza, 2015) when engaging with secondary tasks. This supports the adaption of behaviour at the control level that has also been found in simulator studies (Cnossen et al., 2004; Schömig & Metz, 2013). Yet, there was also a suggestion that drivers strategically plan their engagement with the technological tasks in advance. The notion that drivers' engagement with technology depends on journey type and road infrastructure has also been found in naturalistic driving studies (Tivesten & Dozza, 2015).

There were, however, comments made by drivers that were not included in the development of the original PARRC model from within the literature, which further highlights the importance of assessing the validation of models through application to different sources (Hignett, 2005). This includes an adjustment within the structure of the model formed from the interview data which suggested less of a focus on prioritising goals directly and more reports of adapting their limited resource pool and attributing some resources to determining the priority of the secondary task before lending it any further resources to carry out the task. This highlights the effect that tasks such as text messages have when they draw the drivers' attention towards the device and lead the driver to make a decision to prioritise the driving task or the text task. If the notification did not arise while the vehicle was in motion, they would not be alerted to it and would not need to re-establish their priorities.

Conversely, the connection between the 'behavioural regulation' and 'goal conflict' factors suggests that some drivers were able to control the goals that conflicted with the driving task due to their attitudes, perceptions and behaviour that regulated their engagement with the tasks while driving. This reflects others who suggest the role of driver attitude in their intention to engage with distractions while driving (Walsh et al., 2008; Zhou et al., 2009; Zhou et al., 2012).

The suggestion that drivers' engagement with distractions is largely voluntary (Beanland et al., 2013; Lee, 2014) signifies that they do have an element of control over their behaviour but that they choose to become distracted. The development of the thematic framework and its application of the PARRC model aims to suggest that the drivers' decision to engage is not entirely straightforward and that banning the behaviour through legislation is not the only option. Instead there are numerous factors and actors to consider within the sociotechnical system, which are complexly interconnected in determining the emergence of distraction.

6.4   GENERAL DISCUSSION

Previous research has highlighted the importance of understanding the key underlying causal factors that motivate drivers to engage with technologies and to use these to provide recommendations and countermeasures that limit the adverse effects of driver distraction (e.g. Walsh et al., 2008; Zhou et al., 2009; Young & Lenné, 2010; Atchley et al., 2011; Atchley et al., 2012). Furthermore, the relevance of systems-based measures to counter driver distraction are increasingly understood to be necessary to provide improvements to the safety of the road transport system as a whole (Salmon, McClure & Stanton, 2012; Young & Salmon, 2012; Parnell et al., 2016).

The interviews conducted within this chapter allowed the causal factors that drivers report to influence their likelihood of engaging with technological tasks while driving to be inductively developed into a thematic framework. This framework shows some resemblance to the casual factors taxonomy developed by Stanton and Salmon (2009), while highlighting the importance for the semantic subthemes linked to the driver, the task, the infrastructure and the context surrounding the drivers' interactions with technology. The relevance of the systemic factors to driver distraction suggests how actors outside of the drivers' control may be creating the conditions for distractions to be engaged with. The validity of the framework benefits from an inductive analysis with its generation occurring in a bottom-up fashion, from the drivers' own accounts of their

Creating Conditions for Driver Distraction

behavioural intentions. The detailed methodological process from which the inductive thematic analysis was conducted adds to this validity (Braun & Clarke, 2006).

Application of the thematic framework to the PARRC model in Study 2 suggested how the underlying themes relate to the PARRC factors, validating them with concepts derived from drivers' reports of their own behaviour. Furthermore, exploration of the interconnections between PARRC factors has suggested some similar and some different structural connections within the model. As the original model reflected the empirically tested relationships in the literature, the difference in the interconnections found in this chapter suggest potential gaps in the literature. The inductive analysis has provided factors that drivers themselves report to be important in their decision to engage with technological tasks while driving. This has supported the agenda of previous research that has applied behavioural intention literature to driver distraction (Walsh et al., 2008; Zhou et al. 2009; Zhou et al., 2012). Yet, it has also highlighted the opportunity for future research to explore interacting factors that had not previously been considered in the literature that the PARRC model was grounded in. What has, however, been reinforced is the importance of the interacting elements in the road transport system that create the conditions for drivers to make the decision to engage with technological secondary tasks while driving.

6.4.1 Recommendations to Practise

The adverse implications of using specific technological devices while driving are known (Tsimhoni et al., 2004; Horrey & Wickens, 2006; McCartt et al., 2006; Young, Mitsopoulos-Rubens et al., 2012), yet the facilitating conditions are less acknowledged. By targeting the causal factors of distraction, countermeasures can be developed that focus on the underlying cause of the issue, rather than limiting its effects once engagement has been initiated. The thematic framework developed in Study 1 of this chapter highlights the importance of the wider road transport sociotechnical system and its influence on technology use, including the driver, task, context and road infrastructure. This supports the importance of looking beyond individual focused methods of targeting driver distraction (Salmon et al., 2012; Young & Salmon, 2012; Parnell et al., 2017a).

To determine the importance of systemic actors to issues that are found within sociotechnical systems, the hierarchy of the system can be mapped using the RMF (Rasmussen, 1997). As demonstrated in Chapter 5, the RMF representation of the systems hierarchy is useful in determining how actors interact with each other (Rasmussen, 1997), which can then be used to assess the potential for incident as well as identifying future solutions (e.g. Young & Salmon, 2012). The location of key actors in the hierarchy can facilitate the provision of countermeasures that target elements higher up, to produce widespread change at lower levels (Branford et al., 2009). The use of this hierarchy to review the role that legislation has on the progression of handheld mobile phone-based distractions and distractions from other technologies in Chapter 5 was extended to include an additional two high-level themes, the national and international committees, alongside the original levels; the government, regulators, industrialists, resource providers, end users and the equipment

116 Driver Distraction

and environment. The influence that legislation had on mitigating driver distraction suggested how measures could be taken at each of the levels of the RMF hierarchy to target distraction from in-vehicle technology. The identification of these recommendations to the themes and systemic actors found in the inductive analysis of the drivers' discussions on why they may be likely to engage with distractions in this chapter is illustrated in Figure 6.6. The actors relating to each of the systemic themes and the potential this may hold for the development of future countermeasures to distraction is then discussed.

	Driver	Infrastructure	Task	Context
International Committees		United Nations e.g. The general assembly. ECE Regulations	International standards e.g. ISO, SAE World Health Organisation (WHO)	
National Committees	Culture, Media & Sport Committee	Transport Select Committee	Science & Technology Committee British Standards Organisation	
Government	Department for Culture, Media & Sport	Department for Transport		
Regulators	Media Road Safety Charities	National Road Safety Policy e.g. The Highway Code, Road Traffic Act	UK HMI Design guidelines e.g. BSI code of practise UK Vehicle manufacturing/ design standards	
Industrialists			Aftermarket /portable device manufacturer Telecommunication Association Phone Manufacturer Vehicle Manufacturer	Research Centres
Resource Providers	Driver education/training providers		Aftermarket device designers Driver assistance device designers Phone Designers In-vehicle interface designers	
End-Users	Passenger(s) Other road users			Contacts on media device
Equipment & Environment		Road Type Road features	In-vehicle technology (in-built or portable)	Environmental Conditions

FIGURE 6.6 Systems actors across the RMF that are identified to relate to the four high-level systemic themes identified in the thematic analysis: driver, infrastructure, task and context.

6.4.1.1 Driver

The drivers' attitude towards the use of the technology was found to play an important role in limiting the conflict with the driving goal. Elements of the system that impact on the driver and their attitudes towards technological devices appear across the hierarchy from those directly interacting with the driver, such as passengers or the presence of other road users who may be watching, to higher level actors such as educational providers, the media and road safety charities who can control attitudes in a more top-down manner. A recent road safety campaign by THINK! in the UK with the tag line 'make the glove compartment the phone compartment', guides the driver away from placing the phone goal to conflict with the driving goal (THINK!, 2017). This would also prevent the driver from determining where the task lies in their goal priorities and how they may adapt their behaviour accordingly. Furthermore, the views we have on the use of technology while driving as a society are influenced by national committees who determine the importance of behaviour within national culture. The issue of road safety is a social responsibility that should be shared by the top level of the system (Larsson et al., 2010) and therefore the use of technologies, such as mobile phones while driving need to be portrayed as anti-social when it conflicts with the safe monitoring of the driving task.

6.4.1.2 Infrastructure

As an integral part of the transportation system, road infrastructure is regulated at the international, national and governmental levels with the aim to develop an efficient road transport system (e.g. Department for Transport, 2015b). The interaction of road type with technological engagement has been explored here to identify that drivers do consider the road environment when they decide to use technology while driving, as has also been identified in naturalistic driving studies (Tivesten & Dozza, 2015). On motorways, descriptive themes such as *'just cruising'* and *'consistent'* road layouts suggest that drivers deem the driving task to be less demanding on these roads compared to rural roads where *'poor visibility'* and *'corners'* may conflict with the secondary task. Yet, at the industrial and resource providers level, there are no actors directly influencing which tasks are compatible with different road infrastructures (Salmon, McClure & Stanton, 2012). Future research should determine if certain tasks and technologies are more compatible with certain road environments. For example, interacting and monitoring a satnav may be easier on a motorway, but it has limited use here as there is generally more roadside information and the distance between junctions is greater than on rural or urban roads, which may be more demanding but have a greater requirement to engage with the satnav to navigate through fast changing environments. Tivesten and Dozza (2015) also came to similar conclusions from their naturalistic study and suggested the potential for some road areas where phone use may be regulated, rather than banned.

6.4.1.3 Task

The technological task is associated with a number of systems elements from the very top of the hierarchy to the bottom. The design of technological tasks is influenced by a number of guidelines, standards and criteria that stem from international and national actors, which are then fed down to the individual manufacturers and developers. Yet, there is also a need to represent the views of the end user and apply iterative design procedures that allow for the evaluation of in-built systems usability with respect to the driver and the systems context of use (Harvey et al., 2011b). The drivers interviewed made numerous references to the task features, such as how it may take their eyes off the road or their hands off the wheel under the 'interaction' subtheme, as well as referencing the 'complexity' and 'duration' of the task. This suggests drivers had an understanding of the attentional requirements of the technological tasks. The design standards and guidelines have aimed to inform what is achievable while driving, yet they do not necessarily take into consideration the desire that drivers have to use the technology at the end-user level. Furthermore, it is suggested that different drivers need and desire different information under different contexts (Davidsson & Alm, 2014).

By facilitating the functioning of the technology in the vehicle, the temptation for the driver to engage will endure, and this is particularly true of mobile phones (Nelson et al., 2009). The multifunctionality of smartphones provides extra temptation for the driver to engage with it while driving, and this should be responded to by device developers by limiting functionality while driving. There were numerous comments relating to the notifications received on mobile phone when drivers received texts or phone calls which trigger a response from the driver. The presence of technologies with capabilities to trigger a response that takes the drivers' attention away from the driving task, even momentarily, should be revised by device manufacturers as it forces the end user to assess their priorities which should predominately focus on the driving task and road safety (Lee et al., 2008). The manufacturer Apple has taken steps towards this with their recent update (ios11, released September 2017) that includes a 'do not disturb while driving' mode that can sense when the device is in a moving vehicle and, once prompted by the user, will turn off notifications (Apple, 2017). There is potential for new regulations that target the desires and engagement regulation of technology use to stem from the very top levels of the sociotechnical system and focus on the influences across the levels of the system, not just the driver.

6.4.1.4 Context

The framework highlighted the importance of circumstance in the drivers' decision to engage. There are many complexly interacting factors influencing the use of technology while driving that relate to the road environment, the driver and the task itself. The information that drivers require and desire under different contextual demands is likely to differ (Davidsson & Alm, 2014; Tivesten & Dozza, 2015). The effects of context occur in a predominantly bottom-up fashion within the system hierarchy as they are determined by the interactions with the surrounding environmental conditions. Figure 6.6 shows the lack of high-level actors on context within the sociotechnical system which is reflective of the limited control over the complexly integrated factors that comprise individual circumstances. Road conditions are hard to control as they are influenced by environmental conditions such as time of day, road type and weather conditions. Task conditions relating to the urgency or necessity to interact suggest that drivers assess their priorities as tasks and requirements arise. Yet, determining and setting priorities in advance, or predefining situations

Creating Conditions for Driver Distraction   **119**

where engagement would be more or less necessary could look to control technology engagement at higher levels in the system. Research centres offering facilities to test different contextual factors through the use of driving simulators and highly controlled environments can offer promising insights into the role of context on driver engagement (e.g. Konstantopoulos et al., 2010) and future work should assess the complexly interacting conditions that influence drivers' desire to engage.

6.4.2 EVALUATION AND FUTURE WORK

The thematic framework was developed from the self-reported behaviours of a sample of 30 participants, which is small in contrast to the number of participants that can be recruited from online surveys that facilitate far-reaching recruitment databases. However, the data obtained was much richer, with over 17 h of audio recordings and transcribed data. Furthermore, the sample strived to include drivers of an equal range of age and genders to generate a framework based on a representative sample. While the research focused on UK drivers, the laws relating to technology use in the vehicle are similar across Europe, Australia, New Zealand, Japan and India who specifically ban the use of mobile phones. However, future work should seek to explore how the framework of causal factors may alter with individual and cultural differences, as this may influence the use of technology both inside and outside of the vehicle (e.g. Shinar et al., 2005; Horberry et al., 2006; McEvoy et al., 2006; McEvoy et al., 2007; Young & Lenné, 2010; Young, Rudin-Brown et al, 2012).

An advantage of the thematic framework is its grounding in the self-reported behaviours of drivers, such that the causal factors are directly informed by those who experience them. The use of the interview setting allowed the drivers to openly discuss all factors that they felt influenced their likelihood of engaging. While other studies have explored the drivers' willingness to engage with distractions in simulators (Metz et al., 2011; Schömig & Metz, 2013) and in naturalistic driving settings (Metz et al., 2015; Tivesten & Dozza, 2015), they have looked at the drivers' physical engagement with the task relative to the context of the road environment, rather than enabling the driver to discuss their motivations to engage. The study conducted by Lerner et al. (2008) allowed drivers to discuss their willingness to engage with distractions within focus groups, yet this may have been effected by the social dynamics and biases when discussing behaviours that may be undesirable (Smithson, 2000; Lajunen & Summala, 2003). The drivers' discussions in Study 1 were conducted with confidentiality and anonymity to encourage drivers to reveal their true views. This discussion also allowed for insights into the wider sociotechnical system and its involvement in the issue of driver distraction to be revealed, a concept that was not obtained by the objective measurements in simulator or naturalistic studies (e.g. Metz et al., 2011, 2015; Schömig & Metz, 2013; Tivesten & Dozza, 2015). While it is important to review the drivers' willingness to engage within the context that the behaviour arises, the open discussions with drivers, prompted with different road types and technological tasks, have uncovered a variety of factors that influence their willingness to engage that has not been explored previously. Furthermore, this has provided validation of previous research, as well as proposing the possibility of future explorations of the behaviour and ways to mitigate against it.

6.5 CONCLUSION

This chapter has presented the development of a thematic framework of the causal factors that influence drivers' engagement with different technologies while driving. The use of semi-structured interviews enabled drivers to discuss their likelihood of engaging with a variety of technological tasks across common UK road types. The structure of the interviews allowed all participants to discuss the same tasks and road types while freely generating the key concepts that were important to their perceived likelihood of performing the task. Inductive thematic analysis in Study 1 facilitated the development of a hierarchical framework of causal factors that was driven from the drivers' own interpretations, rather than applying predefined theories. This has shown the influence of systemic actors on the causal factors influencing technology use while driving, highlighting how the road transport system may be creating the conditions for driver distraction to occur. Study 2 showed the relevance of the PARRC factors to the systemic and semantic themes drivers report relate to their engagement in technological devices. This has supported the previous literature in the field from which the PARRC model is grounded in, whilst also suggesting additional concepts of interest. Assessment of the actors impacting on the systemic themes identified from the interviews across the hierarchy of the sociotechnical systems has highlighted future areas for research and countermeasure implementation.

This chapter has shown the utility of semi-structured interviews in the development of the thematic framework that was inductively generated from the drivers' discussions on their likelihood of engaging with the technological devices. The next chapter expands on the findings of the interview study to explore what technologies drivers stated they would be more or less likely to engage with and the influence of the different road types on this. The thematic framework is applied to understand the drivers-stated reasons for their likelihood ratings.

7 What Technologies Do People Use When Driving and Why?

7.1 INTRODUCTION

The previous chapter presented a semi-structured interview study that asked drivers to openly discuss their likelihood of engaging with a range of technological devices. Inductive thematic analysis of the open-ended discussions with drivers generated an in-depth thematic framework that was comprised of hierarchical levels: descriptive, semantic and systemic. This chapter will expand on the findings of the semi-structured interviews to establish what technologies the drivers stated they would be more likely to engage with and why. This will explore the factors that influence the drivers' decision to engage with the technological device, including individual differences, road type and task type. The original methodology detailed in Chapter 6 is expanded on to detail how the interviews allowed the collection of quantitative data alongside the qualitative data and the benefits of this to the findings that were collected. Furthermore, this chapter presents the administration of an online survey that was run alongside the interview study to collect quantitative data from a larger sample of 206 drivers. This enabled a comparison to be made to assess how representative the interview sample was.

7.1.1 FACTORS LINKED TO TECHNOLOGY ENGAGEMENT

There has been a large body of research in the field that has reported age to be a significant factor contributing to drivers' engagement with technological devices (e.g. Lamble et al., 2002; Lerner, 2005; McEvoy et al., 2006; Chen et al., 2016; Pope et al., 2017). This is thought to relate to the relationship between age and access to technology, for example younger drivers have been found to have a higher ownership of mobile phones that has been linked to their increased use while driving in this demographic (Lamble et al., 2002) – although, this effect seems to have decreased in recent years with older adults becoming more accepting of technologies (Mitzner et al., 2010). Yet, older drivers have been found to show more disapproving attitudes towards mobile phone use when driving (Mizenko et al., 2015) and are more in favour of increased restrictions on their use (Lamble et al., 2002). There is also evidence to suggest that older drivers are less able to cope with the increased demand of managing secondary tasks while driving (Alm & Nilsson, 1995; Reed & Green, 1999). Furthermore, Strayer and Drews (2004) found that, while the impact of performing phone-based secondary tasks on driving performance was

121

equivalent in younger and older drivers, the younger drivers' performance when on the phone was the same as the older drivers' when they were not engaging with the phone secondary task. Younger drivers are, however, more likely to underestimate the effect that mobile phones have on their driving behaviour (Tison et al., 2011). McEvoy et al. (2006) found that younger drivers, aged 18–30, were the most likely to self-report engaging with distractions, have a lower perception of risk relating to distracting activities and have had a distraction-related crash, when compared to drivers aged above 30. Although, caution should be heeded when discussing and selecting age categories as recent findings suggest that a middle age category (those between young and old adults) may be similar to the younger drivers in their acceptance and inclination to use technology while driving (Engelberg et al., 2015; Pope et al., 2017).

As a generation of drivers who are accustomed to the use of technologies grow older, the interaction between their cognitive ageing, which is also linked to reduced driving performance, for example Strayer and Drews (2004), and their potential to engage with distractions needs to be considered in future road safety assessments (Pope et al., 2017). Pope et al. (2017) have, however, found a link between the drivers' self-reported engagement with distractions and executive functioning. Their findings suggested that an increased difficulty in executive functioning was related to an increased number of reported engagements with distractions while driving that they suggested could be linked to a lack of ability to inhibit activities (Pope et al., 2017). This provided novel insights into how age may be linked to driver distraction.

The effect of gender and age have also been evidenced with the suggestion that older females are the least likely demographic to engage with mobile phones while driving and younger males the most likely (Pöysti et al, 2005; Lamble et al., 2002).

The impact of distracting tasks has also been linked to environmental conditions and circumstances such as the complexity of the roadway environment (Horberry et al., 2006), manoeuvring different road segments (Lerner & Boyd, 2005) and the curvature of the roadway (Kountouriotis & Merat, 2016). These factors have been found to affect the compensatory mechanisms that drivers employ when engaging with secondary tasks, for example slowing down (Rakauskas et al., 2004; Cnossen et al, 2004), an effect that has been found to be exaggerated with age (Horberry et al., 2006). Road type has also been linked to the structure and content of drivers' situational awareness, with different road environments altering driver perception and behaviour (Walker et al., 2013).

This chapter presents further analysis of the semi-structured interview study detailed in Chapter 6. It aimed to understand both what technological tasks drivers engage with and why. Data from different age groups and genders is compared. The relationship between road type and task type across the different technological devices is explored as there is evidence in the literature to suggest these to be prominent factors implicating driver distraction and technology engagement (e.g. Pöysti et al., 2005; Horberry et al., 2006; McEvoy et al., 2006; Pope et al., 2017). To help mitigate the limitations of a smaller sample size, the interview data is supplemented with data from an online survey.

7.2 METHOD

A trade off was made between the in-depth data analysis of complex open-ended qualitative methods and sample size. Rich interview data could not be collected and analysed in detail from the large samples sizes that are attributed to online surveys. Yet, smaller sample sizes allow for an in-depth understanding of a small group of participants to explore the reasoning behind their behaviour in greater detail (e.g. Dixon et al., 2017). It is, however, useful to know if a small sample are representative of the wider population (Marshall, 1996). To determine if the interview sample were representative, comparisons were made to a large sample of online survey respondents. Although surveys are limited by their inability to infer in-depth knowledge on the drivers' decision-making process, they can be utilised to sample a broader range of UK drivers to determine if the participants' use of technology is representative of a broader range of drivers. As technological developments are rapid, the use of data collected from UK drivers, who are exposed to the same climate of technological development and enforcement of legislation, was sought as an acceptable comparison. The methods used for both the semi-structured interviews and online survey are detailed below. For a more in-depth description of the semi-structured interview methodology, see Chapter 6 (Section 6.2).

7.2.1 Semi-Structured Interview Study

7.2.1.1 Interview Participants

The participants in the semi-structured interview were initially presented in Chapter 6 (Section 6.2.1). The characteristics of the participants are given in more detail here as they relate to the aims and findings of this chapter, see Table 7.1. The 30 licenced UK drivers were recruited across three age categories (18–30 years, 31–49 years and 50–65 years). These age categories were based on the findings from McEvoy et al. (2006) who looked at age effects on engagement with distracting activities using a survey methodology. Confidentiality and anonymity was ensured to allow the participant to talk openly.

7.2.1.2 Interview Procedure

Study 1 in Chapter 6 (Section 6.2) details the procedure involved in conducting the semi-structured interview study. This included the table that was presented to participants and was used to collect the drivers' responses to the Likert scale ratings (Figure 6.1, Chapter 6). Participants were asked to fill in the table by placing a number on the five-point Likert to state their likelihood of engaging in each of the technological tasks across each road type. One on the Likert scale represented 'extremely unlikely' and 5 'extremely likely'. While the Likert ratings were not assessed in Chapter 6, they will be analysed in this chapter alongside the thematic framework that was generated in the previous chapter to further understand the drivers' likelihood of engaging with the different technologies and their reasoning why.

TABLE 7.1
Interview Participant Demographic Information

Demographics		Age Group		
		Younger	Middle	Older
Gender	Female (n)	5	5	5
	Male (n)	5	5	5
Age (years)	Mean (SD)	25.7 (2.41)	36.8 (5.43)	54.9 (2.96)
	Range	22–30	31–47	51–60
Years since passed test (years)	Mean (SD)	6.9 (2.29)	16.9 (6.82)	35.1 (6.73)
Annual mileage (miles)	Mean (SD)	10,850 (4,619)	14,900 (7,445)	11,480 (2,317)
Average hours spent driving a week (hrs)	Mean (SD)	7.5 (3.14)	12.6 (9.27)	9.3 (4.4)

7.2.2 ONLINE SURVEY STUDY

7.2.2.1 Online Survey Participants

A total of 206 participants completed the online survey. They were recruited under the same guidelines as the interview study; they must be frequent drivers, hold a full driving licence and have at least 1 year of experience driving on UK roads. Their demographic information in presented in Table 7.2. The online survey was approved by the institute's ethics committee; Ethical and Research Govenance Online (ERGO) reference: 25219.

TABLE 7.2
Online Survey Participant Demographic Information

Demographics		Age Group		
		Younger	Middle	Older
Gender	Female (n)	35	51	32
	Male (n)	24	28	34
	Rather not say (n)	0	0	2
Age (years)	Mean (SD)	25.7	40.3	57.1
	Range	19–30	32–49	50–75
Years since passed test (years)	Mean (SD)	7.2 (3.2)	19 (7.8)	36.6 (8.6)
Annual mileage (miles)	Mean (SD)	9,449 (9,394)	8,692 (6,114)	9,669 (6,577)
Average hours spent driving a week (hours)	Mean (SD)	6.5 (5.5)	7.3 (4.7)	7.2 (5.1)

Technologies People Use When Driving and Why 125

7.2.2.2 Online Survey Procedure

An online survey was developed using the University of Southampton 'isurvey' platform. Completion of the full survey took approximately 15 min. The survey was designed to explore the drivers' knowledge and interpretation of driver distraction and the laws surrounding it, as well as their driving tendencies. The questions posed to survey respondents correspond to the same questions posed in the interview schedule so that they could be comparable. Importantly, it obtained demographic data from participants and their likelihood ratings for engaging with the same technologies on the same road types as those that were presented to drivers in the interview study. While interview participants were asked to fill in the boxes in Figure 6.1 (Chapter 6) by hand in the online study, drivers selected a number from a drop-down list that related to the same five-point Likert scale as was used in the interviews (1 = extremely unlikely, 5 = extremely likely). The Likert scale ratings could therefore be compared to the interview ratings. Participants were invited to complete the survey via advertisement posters at the university and through the social media outlets of the research group.

7.2.3 DATA ANALYSIS

The online survey provided only quantitative data; the drivers' rated likelihood of engaging with the technological tasks. Data collected from the interviews comprised of both quantitative Likert ratings, suggesting what distractions they would be likely to engage with, and the qualitative reasoning for why, detailed in their verbalised decision process. These two data sets are presented individually within the Results section (Section 7.3.2 presents the quantitative likelihood ratings; Section 7.3.3 presents the qualitative reasoning).

The quantitative ratings from the Likert scales were averaged for comparison to the online survey. Averages were also taken from age and gender categories. The small sample size that is attributed to qualitative methods, such as interviews, limits the statistical analysis that could be conducted to compare these age and gender groups. It also means that the results should be interpreted with caution.

To determine the reasons the drivers gave for their stated likelihood of engaging, the qualitative data generated from the in-depth discussions during the interviews was coded to the thematic framework, as detailed in Chapter 6 using the software tool NVivo 11 (see Figure 6.3). The NVivo 11 software allowed queries to be run on the coded data to review the data in both a quantitative and qualitative manner. It was used to assess the themes of the thematic framework that were referenced across age and gender categories to determine if there were any differences in the reasons drivers from these different categories gave when discussing their likelihood of engaging with the technological devices. A matrix query, run in NVivo 11, allowed the number of references to each of the themes across the age/gender categories to be calculated. It also enabled the data set to be divided to look at the references to the different technological devices individually. The frequency counts from the matrix queries were calculated as percentages to show the percentage of references to the themes between age and gender categories. These are shown in Tables 7.6–7.10. The process is detailed in the flow chart in Figure 7.1.

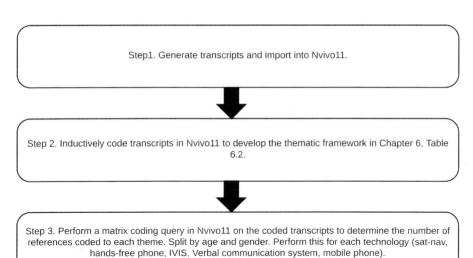

FIGURE 7.1 Flow chart to demonstrate the steps that were taken to determine the percentage of references to the key concepts from the thematic framework.

7.3 RESULTS AND DISCUSSION

7.3.1 Interview and Online Survey Sample Correlation

Comparisons between the interview sample and the online survey sample were made. Due to the unequal sample sizes, a non-parametric Mann-Whitney U test was run to compare the demographics from the two samples. This found no significant differences between the ages (online survey median = 40, interview median = 35, $U = 2{,}783.5$, $p = 0.43$), years since passing driving test (online survey median = 20, interview median = 15, $U = 2{,}726.5$, $p = 0.33$) or hours spent driving a week (online survey median = 7, interview median = 8, $U = 2{,}292$, $p = 0.03$; which was non-significant after applying the Bonferroni correction for multiple tests). This means that the two samples were comprised of a statistically similar range of participants and credits comparisons that can be made from the small interview sample to the larger online survey sample.

Technologies People Use When Driving and Why

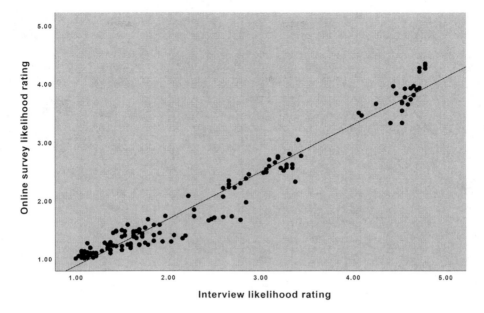

FIGURE 7.2 Scatter plot showing the strong positive relationship between the average reported likelihood of engaging with tasks in the online survey sample and the interview sample.

As the interview sample demographic was found to be representative of the larger interview sample, the relationship between the reported likelihood of engaging with each of the tasks in the online survey sample and the interview sample was explored. This aimed to determine how representative the quantitative findings from the interview sample ($n = 30$) were to a larger sample of UK drivers ($n = 206$). A Spearman's rank-order correlation was run to identify the relationship between the average likelihood of engaging with each of the 22 tasks (Table 6.1, Chapter 6) on each of the 7 road types. This found a strong positive correlation between the average reported likelihood of engaging with the tasks across the road types, between the interview and online survey samples (r_s (152) = 0.95, $p < 0.01$). This is represented in the scatter plot in Figure 7.2. This strong positive correlation indicates that the average reported likelihood of engaging with the tasks by the interview participants is highly consistent with the ratings given by the online survey participants, across both task and road type.

7.3.2 Likelihood Ratings

This section presents the quantitative ratings in response to the Likert scales that participants were asked to complete to inform on their likelihood of engaging with the technological tasks. These are shown for the interview study only. Tables 7.3–7.5 present the Likert scale ratings from the interview study averaged within the younger, middle and older age groups, respectively, and between genders to present the average likelihood ratings for each category in relation to each road type and technological task. These averaged likelihood ratings are coded with shading to represent the ratings on the Likert scale from 'extremely likely' (red) to 'extremely unlikely'

128 Driver Distraction

to engage (green). This generated a heat map of the responses given by participants to highlight what tasks they were more likely to engage with across participants, grouped by age and gender. Heat mapping is a useful tool in presenting patterns and trends in quantitative data in an easily digestible manner, rather than presenting numerical data alone (Bojko, 2009). Although statistical comparisons cannot be conducted due to the small sample sizes, differences between rated engagements can be seen between technology type, task type, road type, age and gender using the shading on the heat maps across Tables 7.3–7.5.

7.3.2.1 Younger Age Category

As can be seen from Table 7.3, younger drivers were more likely to monitor a satnav than enter a destination, although females were even more likely to monitor a navigational app on their phone. Younger drivers were less likely to find a number than answer a call or talk on a hands-free phone. The in-vehicle infotainment systems (IVIS) tasks

TABLE 7.3

Likelihood Likert Scale Ratings for Female and Male Younger Age Category Drivers

| Task | Younger Females | | | | | | | Younger Males | | | | | | |
| | Road Type | | | | | | | | | | | | | |
	Motorway	A/B Road	Urban Road	Rural Road	Residential Road	Junction (Drive Through)	Junction (Stop)	Motorway	A/B Road	Urban Road	Rural Road	Residential Road	Junction (Drive Through)	Junction (Stop)
Navigation system														
Monitor route	4	4	4.3	4.3	4	3.6	3.6	3	3	4.3	4.3	4	3.3	4.6
Enter destination	2	1.3	3	2.3	3.3	1	3.6	1	1	2.3	2	2.3	1	2.6
Hands-free system														
Find number	1.75	1	1.25	1	1.5	0.75	1.75	2	1.3	1.333	1	1.3	1.3	3.6
Answer a call	3.5	3.25	3.25	2.75	3.5	2.75	3.75	5	5	5	4.3	5	2.6	5
Talk to other	3.75	3.75	3.5	2.75	3.75	2.75	3	5	5	4.3	4.3	5	5	5
In-vehicle system														
Change climate control	4.75	4.75	4.5	4	4.75	1.75	4.25	5	5	4.8	4.6	4.8	3.8	5
Change song/radio station	5	4	4.75	4.5	4.75	2.25	4.25	4.8	4.6	3.8	4.2	3.6	3	4.8
Adjust volume	5	5	5	4.5	5	3.25	5	5	5	4.8	5	4.8	4.4	5
Listen to music	5	5	5	4.5	4.75	4.75	5	5	5	4.8	4.8	4.8	4.8	4.8
Verbal comms.	4.5	4.25	4.5	3.75	4.5	2.5	4.75	5	5	4	4	4.3	4	5
Mobile phone/Portable device														
Enter destination (nav app)	2.75	2	1.5	1	1.25	1	3.75	2	1.4	1	1.2	1	1	2.8
Monitor (nav app)	5	5	5	5	5	4	4	2.8	2.8	3	3	3	3	3.6
Write/send a text	2	1.5	1.25	1	1.25	1	2.75	2.4	2.2	1.4	1.2	1.4	1.2	2.6
Read a text	3.25	2.75	3	2.5	3	1	3.5	2.6	2.6	2	2	2.2	2	3.4
Answer phone call	3	2.5	2	1.75	2	1	2	2.8	2.8	2	2.4	2.4	2.2	2.6
Talk on the phone	2	1.5	1.75	1.25	1.75	1	1.75	2.8	2.8	2.2	2.2	2.6	2.2	2.6
Enter/find a number	2	1.5	1.5	1.25	1.5	1	2.75	2	1.8	1.4	1.2	1.4	1.2	2.6
Change song/audio track	3.25	2.75	2.75	2	2.5	1	3.25	2.2	2	1.8	1.6	1.8	1.6	2.8
Use voice assist features	2.75	2.25	2.5	2.25	2.5	1	2.75	1.4	1.4	1.4	1.4	1.4	1.4	1.6
Take a photo	1	1	1	1	1	1	1.5	1.6	1.2	1	1	1	1	1.8
Use social media apps	1.25	1.25	1.25	1.25	1.25	1.25	2	1	1	1	1	1	1	1.6
Check your email	2	1.75	1.25	1	1	1	1.5	1.4	1.4	1	1	1.2	1	2.2

Dark Grey, extremely likely to engage; White, extremely unlikely to engage.

Technologies People Use When Driving and Why

were reported to be the most likely to be engaged with, although they were reported to be less likely when driving through a junction. Younger drivers were more likely to report engaging with mobile phone tasks when stopped at junctions and least likely when driving through a junction. They also reported themselves to be moderately likely to engage with a number of phone-based tasks while driving across different road types.

7.3.2.2 Middle Age Category

Table 7.4 shows drivers in the middle age category were more likely to monitor the destination on a navigation system than enter a destination, with females slightly more likely than males. Middle aged female drivers were more likely to answer a call and talk on a hands-free phone than find a number on the device. Males were more equal in their ratings of finding a number and answering a call. Most drivers in the middle age category were likely to engage in tasks associated with the IVIS, apart from the verbal communication systems, which drivers were unlikely to engage

TABLE 7.4
Likelihood Likert Scale Ratings for the Female and Male Middle Age Category Drivers

Task	Middle Females							Middle Males						
	Motorway	A/B Road	Urban Road	Rural Road	Residential Road	Junction (Drive Through)	Junction (Stop)	Motorway	A/B Road	Urban Road	Rural Road	Residential Road	Junction (Drive Through)	Junction (Stop)
Navigation system														
Monitor route	3.75	4.5	5	4.25	4.25	4.25	3.75	2.75	3	3.75	2.75	3.5	1.5	3
Enter destination	2.25	2.25	2	2.25	1.5	1.5	3.25	2.25	2	2	1.5	2	1	3
Hands-free system														
Find number from address book	2	2	2	2	2	2	2	4	3.5	3	2	3	1	4
Answer a call	3.667	3.667	4.6	4.6	5	3.3	5	4	3.5	3.5	3	3	2.5	5
Talk to other	3.667	3.333	4.3	4.3	4.6	4	5	4	4.5	5	3.5	4.5	4.5	5
In-vehicle system														
Change climate control	4.8	4.8	4.8	4.8	4.8	2.8	5	5	5	5	4.8	4.8	3.2	5
Change song/radio station	4.6	4.6	4.6	4.2	4.4	3	5	5	5	4.8	4	4.6	2.6	5
Adjust volume	4.8	4.8	4.8	4.8	4.8	3	5	5	5	5	5	5	3.6	5
Listen to music	5	5	5	5	5	4.2	5	5	5	5	5	5	5	5
Verbal comms.	1	1	1	1	1	1	1	2.3	2.3	2.333	2.333	2.3	1.3	2.3
Mobile phone/Portable device														
Enter destination (nav app)	1.6	1.6	1.6	1.6	1.6	1	2.4	2.8	2.8	2.6	2.4	2	2	3.4
Monitor (nav app)	2.4	2.4	2.6	2.4	2.4	2	2.8	2.8	2.8	3.2	3	3	2.4	3.4
Write/send a text	1	1	1	1.2	1	1	1.8	1.8	1.6	1.4	1.2	1.2	1.2	3.2
Read a text	1.4	1.8	2.2	1.6	1.6	1	4	3.4	3.2	3	3	3	2.2	4.4
Answer phone call	1.8	1	1.8	1	1	1	1	3	2.8	2.6	2.6	2.6	1.2	4
Talk on the phone	1.8	1	1.8	1	1	1	1	2.8	2.6	2.6	2.6	2.4	1.8	3.8
Enter/find a number	1.8	1.8	1.6	1.8	2	1.6	2.8	2.8	2.6	2.4	2	2.4	1.6	4
Change song/audio track	1.4	1.4	1.4	1.6	1.6	1	1.6	2.4	2.4	2.2	2.2	1.8	1.2	2.6
Use voice assist features	1.8	1.8	1.6	1.8	1.4	1	1.8	2.2	2.2	2.2	1.8	2.2	1.2	1.6
Take a photo	1.6	1.6	1.6	1.6	1.6	1	1.2	1	1	1	1	1	1	2.2
Use social media apps	1	1	1	1	1	1	2	1.2	1.2	1.2	1.2	1.2	1	2.4
Check your email	1	1	1	1	1	1	2	2.2	2.2	2.2	2.2	1.8	1.8	3.6

Dark Grey, extremely likely to engage; White, extremely unlikely to engage.

130 Driver Distraction

with. Males were more likely to interact with a mobile phone than females, yet when stopped at a junction, interactions were rated to be more likely by both genders.

7.3.2.3 Older Age Category

As can be seen from Table 7.5, older drivers were far more likely to monitor a route on a satnav than enter a destination. No older females from the interview sample had access to a hands-free system so there are no results for this technology. Older males, however, were likely to answer a call and talk on a hands-free system. Again, the IVIS task was the most likely to be reported, although similar to the middle age category, they were a lot less likely to use verbal communication systems. The older females were extremely unlikely to use their mobile phone, although they were slightly more inclined to read a text while on a motorway or A/B road. Older males were more likely to use their mobile phone than the females, but they were still less likely than all other age groups.

TABLE 7.5

Likelihood Likert Scale Ratings for the Female and Male Older Drivers Age Category

Task	Older Females							Older Males						
	Road Type													
	Motorway	A/B Road	Urban Road	Rural Road	Residential Road	Junction (Drive Through)	Junction (Stop)	Motorway	A/B Road	Urban Road	Rural Road	Residential Road	Junction (Drive Through)	Junction (Stop)
Navigation system														
Monitor route	3	3	2.75	2.5	2.5	2.75	3.25	3.8	4	4.2	4.2	4.2	2.8	4
Enter destination	1	1.25	1.25	1	1	1	1	1.4	1.4	1.2	1.2	1	1	2
Hands-free system														
Find number from address book	-	-	-	-	-	-	-	1	1	1	1	1	1	2
Answer a call	-	-	-	-	-	-	-	3.5	3.5	3	3	3	3	3.5
Talk to other	-	-	-	-	-	-	-	4	4	4	4	4	4	4
In-vehicle system														
Change climate control	3.8	3.8	3.6	3.6	3.6	2.8	3.6	5	5	4.8	4.8	4.6	3	5
Change song/radio station	4.4	4.2	3.8	3.6	3.4	2.6	3.4	4.6	4.6	3.6	3.6	3.6	2	4.4
Adjust volume	4	4	4	4	3.8	3.2	3.8	5	5	4.8	4.8	4.2	3.2	5
Listen to music	4.8	4.8	4.8	4.8	4.8	4.8	4.8	4.4	4.4	4.4	4.4	4.2	4.4	4.4
Verbally comms.	2.5	2.5	2.5	2.5	2.5	2.5	2.5	2.3	2.3	1.667	2	1.3	1	2.3
Mobile phone/Portable device														
Enter destination (nav app)	1	1	1	1	1	1	1	1	1	1	1	1.2	1	1.6
Monitor (nav app)	1	1	1	1	1	1	1	2.8	2.8	2.6	2.6	2.2	1.8	2.6
Write/send a text	1	1	1	1	1	1	1	1	1	1	1	1	1	1.4
Read a text	1.4	1.2	1	1	1	1	1	1.8	1.8	1.6	1.6	1.6	1.2	2
Answer phone call	1	1	1	1	1	1	1	1.6	1.6	1.4	1.4	1.4	1.4	1.6
Talk on the phone	1	1	1	1	1	1	1	1.6	1.6	1.4	1.4	1.4	1.4	1.6
Enter/find a number	1	1	1	1	1	1	1	1	1	1	1	1	1	1.2
Change song/audio track	1	1	1	1	1	1	1	1.4	1.4	1.2	1.2	1	1	1.2
Use voice assist features	1	1	1	1	1	1	1	1.4	1.4	1.2	1.2	1.4	1.2	1.4
Take a photo	1	1	1	1	1	1	1	1	1	1	1	1	1	1
Use social media apps	1	1	1	1	1	1	1	1	1	1	1	1	1	1
Check your email	1	1	1	1	1	1	1	1	1	1	1	1	1	1

Dark Grey, extremely likely to engage; White, extremely unlikely to engage.

Technologies People Use When Driving and Why

7.3.3 LIKELIHOOD REASONING

The insights from the drivers' open-ended discussions on why they may be more or less likely to engage with the technologies are presented in the following sections. Tables show the percentage ratings that were calculated using the matrix query function in NVivo 11, as detailed in Figure 7.1. The total number of references to the four main systems themes (Driver, Task, Context, Infrastructure) is aggregated from their comprising subthemes listed below them. The percentages of references to each of the key themes in relation to each of the technologies are presented across the age/gender categories in the Tables 7.6–7.10. Again, heat-map shading has been applied to highlight the different percentage of references to the themes; red indicates a higher frequency of references, green indicates a lower frequency of references to the theme.

7.3.3.1 Satnav

The desire to use a satnav was discussed, with most participants stating that they would only use the satnav if they did not know where they were going, although two participants commented that they use the satnav if they do know where they are going as it gives them traffic information and time until arrival. There were some differences in the likelihood of engaging with tasks on the specific satnav device, with the young more likely to monitor the navigation app on their phone than use a satnav device (Table 7.3). A shift in the use of smartphone navigation by young people in recent years has been found elsewhere in the literature, which has made a link to the heightened confidence young people have in using the technology (Speake, 2015). Yet, when technologies malfunction, it can lead to a feeling of loss of control and difficulties in trying to reconnect with the surrounding environment (Speake, 2015). Young drivers were, however, less likely to enter a destination on the phone navigation app than to enter a destination on the satnav device. The laws banning mobile phone interaction while driving may indicate why drivers report a reduced likelihood for this task on the phone device. Discussion related to the illegality theme and mobile phones are explored later (Section 7.3.3.7), but in relation to satnav use, no references by any interviewee were made to the law surrounding satnav use. This could be because there are no specific laws relating to the drivers' use of the technology, rather there are generic laws relating to driving with 'due care and attention' (see Chapter 5; Parnell et al., 2017a).

The discussion with the drivers on their likelihood of engaging with the satnav tasks was, for the majority of participants, focused on the road infrastructure. The use of the satnav to navigate the road environment is likely to be the reason for this, the only exception was the older males (Table 7.6). Older males were slightly more concerned with the features of the tasks themselves and largely their interaction with it. For example, one older male commented that they wouldn't enter a destination into the satnav *"because you've got to check you've spelt it right and then you've got to press enter"* (participant 17). This supports previous findings of a reduced ability to cope with managing the increased demand of secondary tasks in older drivers (Alm & Nilsson, 1995; Reed & Green, 1999), adding that older drivers are aware of their difficulties.

132   Driver Distraction

TABLE 7.6

Percentage of References to Each Theme When Drivers Discussed Their Likelihood to Engage with the Satnav Tasks While Driving

Themes	Young		Middle		Older	
	Female (%)	Male (%)	Female (%)	Male (%)	Female (%)	Male (%)
Driver	6.7	0.0	16.3	2.8	15.0	10.5
Attitude towards task	0.0	0.0	2.0	0.0	12.5	5.3
Influence of others	0.0	0.0	2.0	0.0	0.0	0.0
Tendency	6.7	0.0	8.2	2.8	0.0	2.6
View of self-behaviour	0.0	0.0	4.1	0.0	2.5	2.6
Task	33.3	42.5	28.6	22.2	25.0	44.7
Ability to complete	0.0	0.0	2.0	0.0	10.0	2.6
Complexity	6.7	2.5	0.0	2.8	2.5	5.3
Desirability	6.7	15.0	8.2	8.3	2.5	7.9
Duration	0.0	5.0	0.0	0.0	0.0	5.3
Engagement regulation	6.7	12.5	10.2	2.8	5.0	7.9
Interaction	13.3	7.5	8.2	8.3	5.0	15.8
Context	6.7	2.5	4.1	11.1	7.5	7.9
Journey Context	6.7	0.0	0.0	5.6	7.5	0.0
Road context	0.0	0.0	0.0	0.0	0.0	2.6
Task Context	0.0	2.5	4.1	5.6	0.0	5.3
Infrastructure	53.3	55.0	51.0	63.9	52.5	36.8
Illegality	0.0	0.0	0.0	0.0	0.0	0.0
Road layout	6.7	7.5	18.4	30.6	12.5	10.5
Road related behaviour	6.7	10.0	4.1	5.6	2.5	2.6
Task-road relationship	33.3	35.0	18.4	19.4	30.0	21.1
Perceptions of sur. env.	6.7	2.5	10.2	8.3	7.5	2.6

Dark Grey, extremely likely to engage; White, extremely unlikely to engage.

The high number of references to the 'task-road relationship' theme highlights how drivers change their use of the satnav depending on the road environment. For example *"when I'm on the motorway I'll sort of look to see when the turning off is and then I just sort of relax after that. I don't really stare at it"* (participant 13) and *"Residential I would probably be quite likely to do that, again you are crawling along and that is often when you need the sat nav - at the end of the journey"* (participant 1). Previous research has suggested that satnavs are changing the way in which we navigate (Axon et al., 2012), the egocentric display, in contrast to the objectivity of traditional maps, allow users to only view what is relevant to their journey (Meng, 2004). This has been found to increase the drivers' feeling of control (Speake, 2015). The findings from this study suggest that drivers are utilising the egocentric displays to only engage with them at points within the road environment that they deem to be relevant. The use of voice commands alongside visual displays has been recommended in the design of navigation systems (Green et al., 1995). While auditory cues limit the drivers' diversion of visual attention away from the driving task, drivers have shown a preference for visual navigational displays that allow them to view their location when and as they require it (Srinivasan & Jovanis, 1997; Burnett, 2000).

7.3.3.2   Hands-Free Phone

As shown in Table 7.5, no females in the older category owned a hands-free phone, so they could not be included in this comparison. Across the other age groups,

Technologies People Use When Driving and Why 133

there was a clear distinction between the task 'find a number' and the other hands-free tasks: 'answer a call' and 'talk to other' (Tables 7.3 and 7.4). This suggests that drivers were less likely to initiate a call by finding a contact than responding to a phone call and maintaining a conversation. This supports findings from the literature on the distinction between initiating and responding to phone communications, with the desire to respond higher (Atchley et al., 2011; Waddell & Weiner, 2014; Nemme & White, 2010). The exception was the males in the middle age category, where this distinction is less clear and answering a call is as likely as initiating one. Drivers mostly reported the same likelihood of answering a call as talking on the phone, as one leads to another. Yet, at some points drivers stated that they may delay answering the phone until they were at a point where they felt confident to talk, or they may answer a call but pause the conversation to delay speaking until they are confident to do so. For example, *"So I think answering a call is distinctly different to just talking, for the reason that on rural roads if someone had called me then I'd check to make sure the road is all right before I went to answer the call again. As opposed to if you were on a motorway you've got a lot more time to just be able to look at it and then flick back and then pick it up, so you wouldn't have that kind of delay"* (participant 4). This supports evidence that drivers are able to adapt their behaviour in line with their perceived demands of the road environment (e.g. Rakauskas et al., 2004; Cnossen et al., 2004; Parnell et al., 2016).

Previous research has linked the willingness to engage with phone communications to social factors (e.g. Lerner et al., 2005). The interview responses coded to the thematic framework, however, suggested less of an effect of social pressure in relation to hands-free phone use. The younger drivers, and the middle age category males, cited factors relating to the infrastructure to influence their likelihood of engaging with hands-free tasks (Table 7.7). For the middle aged males, these comments were largely focused on the road-related behaviour theme and the task-road relationship, with engagement in the task differing on the features of the road environment that may influence their ability to interact with the task, for example *"You see I actually think urban road, answering a call I'd probably be more likely to than on a major road. I think that might just be because of the driving speeds"* (participant 7). Younger drivers, who are arguably the most swayed by social pressures in relation to risk taking (Gardner & Steinberg, 2005), cited factors relating to the infrastructure the most in relation to their willingness to engage, for example *"On a motorway, you only really have to know when to stop so as long as I know that a car is very, very far ahead of me I can do what I like when that's happening"* (participant 13). The absence of comments relating to the 'influence of others' on hands-free engagement suggest that social factors are less important to young drivers than was suggested for mobile phone communication by Lerner et al. (2005).

In line with previous findings related to handheld phones while driving that suggest older drivers hold stronger disapproving attitudes (Mizenko et al., 2015), the older males were the most likely to report an attitude to the task which influenced their likelihood of engaging, one simply responded, *"I think it's inherently unsafe"* (participant 8).

TABLE 7.7

Percentage of References to Each Theme When Drivers Discussed Their Likelihood to Engage with Hands-Free Phone Tasks While Driving

	Young		Middle		Older	
Themes	Female (%)	Male (%)	Female (%)	Male (%)	Female (%)	Male (%)
Driver	2.4	5.4	14.3	14.8	-	25.0
Attitude towards task	2.4	5.4	0.0	3.7	-	12.5
Influence of others	0.0	0.0	0.0	0.0	-	6.3
Tendency	0.0	0.0	14.3	11.1	-	6.3
View of self-behaviour	0.0	0.0	0.0	0.0	-	0.0
Task	24.4	29.7	50.0	18.5	-	18.8
Ability to complete	7.3	0.0	7.1	7.4	-	12.5
Complexity	2.4	8.1	7.1	3.7	-	0.0
Desirability	4.9	5.4	0.0	3.7	-	6.3
Duration	0.0	0.0	0.0	0.0	-	0.0
Engagement regulation	2.4	8.1	14.3	0.0	-	0.0
Interaction	7.3	8.1	21.4	3.7	-	0.0
Context	17.1	13.5	21.4	3.7	-	37.5
Journey Context	2.4	2.7	0.0	0.0	-	0.0
Road context	12.2	5.4	14.3	3.7	-	0.0
Task Context	2.4	5.4	7.1	0.0	-	37.5
Infrastructure	56.1	51.4	14.3	63.0	-	18.8
Illegality	0.0	0.0	0.0	0.0	-	0.0
Road layout	2.4	5.4	0.0	18.5	-	6.3
Road related behaviour	4.9	8.1	0.0	25.9	-	6.3
Task-road relationship	24.4	18.9	14.3	14.8	-	6.3
Perceptions of sur. env.	24.4	18.9	0.0	3.7	-	0.0

Dark Grey, extremely likely to engage; White, extremely unlikely to engage.

7.3.3.3 In-Vehicle Infotainment Systems

The IVIS tasks were the most highly rated tasks to be engaged with across all age groups, with most drivers rating themselves to be highly likely to use the system across all road types, except when driving through junctions (Tables 7.3–7.5). Similarly, Lansdown (2012) found that the in-car entertainment system was one of the more frequent activities undertaken by drivers, suggesting that this may be due to the low level of risk associated with these tasks becoming distractors. This is evidenced by one participant from the younger male category who stated the following: *"I feel because it's part of the car, I feel as if it's okay…it's as if the car is sort of saying it's okay to do this so you do just sort of click it and change it up so I'd say a five for all of them"* (participant 13). They are making the connection that they felt that the controls located on the IVIS were fine to use because manufacturers have put them in the vehicle purposefully. This emphasises the responsibility that manufacturers have when placing technologies into vehicles to ensure that they will not result in unnecessary distraction from the driving task, as drivers are encouraged to interact with features that are built-in to the vehicle.

Indeed, the high ratings of engagement relating to IVIS tasks can be attributed to the comments by participants on the features of the tasks themselves (Table 7.8). There were multiple comments on the ease of the tasks ('complexity' and 'ability to complete' themes, Table 7.8), with pre-programmed buttons for the radio, and

Technologies People Use When Driving and Why

TABLE 7.8

Percentage of References to Each Theme When Drivers Discussed Their Likelihood to Engage with the IVIS Tasks While Driving

Themes	Young Female (%)	Young Male (%)	Middle Female (%)	Middle Male (%)	Older Female (%)	Older Male (%)
Driver	15.0	10.9	15.6	15.7	13.6	21.4
Attitude towards task	8.8	5.4	0.0	7.8	5.1	11.4
Influence of others	0.0	0.0	0.0	0.0	0.0	0.0
Tendency	5.0	5.4	14.1	7.8	8.5	10.0
View of self-behaviour	1.3	0.0	1.6	0.0	0.0	0.0
Task	42.5	54.3	57.8	54.9	44.1	42.9
Ability to complete	2.5	13.0	6.3	9.8	6.8	10.0
Complexity	6.3	6.5	6.3	7.8	6.8	11.4
Desirability	6.3	9.8	14.1	7.8	3.4	4.3
Duration	2.5	0.0	1.6	0.0	1.7	0.0
Engagement regulation	1.3	2.2	9.4	7.8	10.2	4.3
Interaction	23.8	22.8	20.3	21.6	15.3	12.9
Context	12.5	8.7	4.7	7.8	11.9	8.6
Journey Context	0.0	2.2	0.0	0.0	3.4	0.0
Road context	10.0	6.5	3.1	5.9	6.8	7.1
Task Context	2.5	0.0	1.6	2.0	1.7	1.4
Infrastructure	30.0	26.1	21.9	21.6	30.5	27.1
Illegality	0.0	0.0	0.0	0.0	0.0	0.0
Road layout	3.8	3.3	6.3	3.9	1.7	2.9
Road related behaviour	3.8	0.0	3.1	5.9	8.5	2.9
Task-road relationship	13.8	15.2	6.3	5.9	13.6	11.4
Perceptions of sur. env.	8.8	7.6	6.3	5.9	6.8	10.0

Dark Grey, extremely likely to engage; White, extremely unlikely to engage.

steering wheel controls increasing the likelihood that drivers reported for engaging with the IVIS. Many commented that the use of the steering wheel buttons to control tasks, such as changing songs or volume, meant that they could perform the task without looking away from the road, so they were happy to engage in the task most of the time. For example, *"Adjust volume, again I am very likely to. Good news on this I can do it at the steering wheel so I don't need to look at that, I can just feel it and do it"* (participant 25).

The voice command aspect of the IVIS produced markedly different results to the other physical interactions with the system, therefore this was explored separately in more detail.

7.3.3.4 Voice Command System

A large distinction in the use of these systems was found between those in the younger age category, who were more likely to use voice command systems across road types, and the middle and older age categories who were much less likely to use the system anywhere (Tables 7.3–7.5). For the younger group, verbal interactions with the in-vehicle system did not differ to their physical interactions with the IVIS, such as changing the radio or adjusting the volume (Table 7.3). Analysis of the reasoning given by participants highlighted the large number of references to the features of the tasks in relation to the voice system (Table 7.9). Although all participants reported that features of the task influenced their decision, the younger group were

TABLE 7.9

Percentage of References to Each Theme When Drivers Discussed Their Likelihood to Engage with the Voice Command Tasks While Driving

Themes	Young Female (%)	Young Male (%)	Middle Female (%)	Middle Male (%)	Older Female (%)	Older Male (%)
Driver	9.1	12.5	12.5	10.5	11.1	15.8
Attitude towards task	0.0	6.3	0.0	0.0	5.6	10.5
Influence of others	0.0	0.0	0.0	0.0	0.0	0.0
Tendency	4.5	3.1	8.3	5.3	0.0	5.3
View of self-behaviour	4.5	3.1	4.2	5.3	5.6	0.0
Task	82.0	84.4	75.0	84.2	88.9	78.9
Ability to complete	9.1	6.3	0.0	0.0	11.1	0.0
Complexity	9.1	3.1	4.2	5.3	0.0	10.5
Desirability	40.9	43.8	25.0	36.8	16.7	21.1
Duration	0.0	3.1	0.0	0.0	0.0	0.0
Engagement regulation	22.7	12.5	33.3	36.8	50.0	36.8
Interaction	0.0	15.6	12.5	5.3	11.1	10.5
Context	4.5	3.1	4.2	0.0	0.0	5.3
Journey Context	0.0	0.0	0.0	0.0	0.0	0.0
Road context	4.5	3.1	4.2	0.0	0.0	5.3
Task Context	0.0	0.0	0.0	0.0	0.0	0.0
Infrastructure	4.5	0.0	8.3	5.3	0.0	0.0
Illegality	0.0	0.0	0.0	0.0	0.0	0.0
Road layout	0.0	0.0	0.0	0.0	0.0	0.0
Road related behaviour	4.5	0.0	4.2	0.0	0.0	0.0
Task-road relationship	0.0	0.0	4.2	5.3	0.0	0.0
Perceptions of sur. env.	0.0	0.0	0.0	0.0	0.0	0.0

Dark Grey, extremely likely to engage; White, extremely unlikely to engage.

not deterred by any failures of the system and would 'have a go' at using it anyway, whereas in the middle and older categories, they were more likely to be put off using the voice system as they felt it would not work properly. A young female commented: *"yeah I would probably do that any time as it is just – "just talking". Although when it doesn't work and you get annoyed then that is probably more detrimental than anything else"* (participant 27). Similarly, a young male stated that *"if I get it wrong or don't complete it, the consequences are minor or irrelevant and I don't have to take my eyes off the road"* (participant 5). Whereas an older male commented *"I've been in a car where my friend's been trying to do it and it's a bit of a joke trying to get the thing to understand you, so I think … I think my car does it, but I haven't attempted to get it working so extremely unlikely"* (participant 8). This suggests that the unreliability of the system prevented them from even attempting to engage with it. This supports previous findings by Lee et al. (2015), who also found that age was inversely related to self-reported technology experience in in-car voice controlled systems and that older drivers were more likely to spend longer trying to perform voice control tasks, finding them difficult to perform and requiring extra prompts. It was therefore evident that voice command technology still has some way to go to gain the trust of the consumer, despite much technological advancement since their conception (Furui, 2010). Further work is required to determine the use of voice command systems from a larger population.

7.3.3.5 Mobile Phone

Similar to the hands-free phone tasks, the findings from the mobile phone ratings suggest some differences between the initiation behaviours of a phone call, that is finding a number, and answering a phone call. This distinction was clear in the younger male and female categories (Table 7.3). The middle age category males showed little differentiation between the two behaviours, suggesting they would be somewhat likely to engage in both initiation and response behaviours across road types, especially when stopped at a junction (Table 7.4). The older age group reported that they were generally 'extremely unlikely' to use their mobile phones, although the males suggested that they may be more likely than the females to read texts and answer/talk on the phone (Table 7.5).

The age difference observed between the older and the younger age groups may be due to the participants in the older category expressing more difficulty in performing the tasks on the phone, for example: *"Send a text on a hand-held device? I can't do that when I'm stood still in a car park, let alone driving along, so no, no, no, no"* (participant 3). Although another older male did state that they have read a text on the phone, they also report the adverse effects of doing so, for example *"Reading a text also seems a bit dangerous to me. But you can... driving, I have done it, but not... I have a... it's a big dodgy"* (participant 8). Furthermore, the older age category referenced their attitude towards the mobile phone task considerably more than those in the other age categories (Table 7.10). The older participants tended to be stronger minded in their views on the use of mobiles phones, leading to the lowest scores for this device (Table 7.5). One older female commented that the use of her phone related to when she wanted to engage with it, rather than her use being dictated by the person trying to get hold of her: *"the phone is for me, not them. It is having just a bit of attitude!"* (participant 11). Another stated: *"Because nothing is that pressing that it is worth getting yourself killed for, is it? That is the way I look at it. Or killing other people"* (participant 12). This supports previous research stating that older drivers are more disapproving of mobile phone use when driving than younger drivers and are less likely to use their phone behind the wheel (Mizenko et al., 2015; Lamble et al., 2002). Furthermore, it was the older females who voiced more attitude references to mobile phone use and were least likely to engage in any mobile phone task (Table 7.5). Pöysti et al. (2005) also found older females to be the least likely to engage with mobile phones while driving and younger males the most likely.

Conversely, the younger participants stated that they would more readily read a text message as they pop up on the phone screen, for example *"Yes, I would read a text. It just pops up on your phone. It's impossible not to really"* (participant 13). *"I'd definitely see who it was from probably pretty much straight away"* (participant 4). The younger group also report the ease of multitasking with the mobile phone, in contrast with the comments of the older drivers, for example *"If it's just a couple of words then you can read it quite easily"* (participant 4). This supports previous findings that they may underestimate the adverse effects of the task on their driving performance (Tison et al., 2011). Yet, again rather than being influenced by social influencers (Atchley et al., 2011; Lerner et al., 2005), comments from the interview transcripts in relation to the use of mobile phones while driving were heavily coded to 'task' themes in the thematic framework.

TABLE 7.10

Percentage of References to Each Theme when Drivers Discussed their Likelihood to Engage with Mobile Phone Tasks While Driving

Themes	Young		Middle		Older	
	Female (%)	Male (%)	Female (%)	Male (%)	Female (%)	Male (%)
Driver	18.1	22.0	21.4	21.9	34.4	35.6
Attitude towards task	9.3	10.3	6.3	10.6	27.1	21.1
Influence of others	3.1	2.3	1.9	0.7	2.9	1.1
Tendency	5.2	8.0	10.7	6.6	2.9	11.1
View of self-behaviour	0.5	1.4	2.5	4.0	1.4	2.2
Task	44.7	52.1	52.9	47.7	52.8	45.5
Ability to complete	4.7	2.8	6.3	4.0	7.1	5.6
Complexity	4.7	2.8	1.9	3.3	0.0	4.4
Desirability	13.0	18.8	14.5	13.2	8.6	13.3
Duration	2.1	6.6	1.9	0.7	1.4	0.0
Engagement regulation	5.7	7.0	11.3	11.9	15.7	13.3
Interaction	14.5	14.1	17.0	14.6	20.0	8.9
Context	11.8	7.2	12.2	12.8	5.9	11.2
Journey Context	1.6	1.4	0.6	0.7	1.4	0.0
Road context	2.6	1.4	1.9	1.3	0.0	1.1
Task Context	7.3	4.2	9.4	10.6	4.3	10.0
Infrastructure	26.0	18.9	13.8	18.0	7.1	7.7
Illegality	2.1	1.9	3.1	0.7	2.9	1.1
Road layout	7.3	0.5	1.3	6.0	1.4	1.1
Road related behaviour	2.6	3.8	1.9	0.7	1.4	1.1
Task-road relationship	8.3	6.1	4.4	8.6	0.0	3.3
Perceptions of sur. env.	5.7	6.6	3.1	2.0	1.4	1.1

Dark Grey, extremely likely to engage; White, extremely unlikely to engage.

There were numerous discussions engaged by participants on the location of their phone while driving, with drivers stating the importance of this on their interaction with it. Some participants kept their phone in their handbag, others in their pocket while some had phone holders to attach their phone to the dashboard. These were given as reasons for why they may be more or less likely to interact with the phone while driving, for example: *"Normally I would leave my phone in my handbag, it automatically connects to the car, so then it would only be if I was stuck in traffic that I would probably get it out and see if there is anything on there"* (participant 22). This is in contrast to another participant *"So it's clipped to one of my air vents right next to my – in the centre of my car, right next to my steering wheel, so I can glance over and see it better"* (participant 7). The location of the phone had particular relevance to the task of 'reading a text'. This was the one phone-related task which all participants stated they were more likely to perform than others. Even those in the older age category stated that they would perform this task under some circumstances, with the older females giving the only rating greater than 1 ('extremely unlikely', the lowest score on the scale) to the task of reading a text on a motorway or a major road. This highlights the importance of considering where drivers are placing their portable devices in the vehicle.

Technologies People Use When Driving and Why

A new UK road safety campaign has recently started to consider phone location with the tag line 'make the glove compartment the phone compartment', guiding the driver away from placing the phone in conflict with the driving task (THINK!, 2017). Yet, this may conflict with the other tasks that drivers use their phone for while driving, such as navigation and music systems. Lerner et al. (2005) highlighted the importance of anticipation and preparation for technology use when driving, such as placing technology like Bluetooth earpieces in easy-to-reach locations while driving to limit distraction.

7.3.3.6 Road Type

Reviewing the heat maps of the Likert scale ratings in Tables 7.3–7.5 suggests a greater difference in likelihood ratings between task type than road type, which is consistent with Lerner et al. (2005). Yet, Lerner et al. (2005) found the effect of road type was more pronounced when they looked at the effect of performing manoeuvres while driving. For example willingness decreased when exiting and entering a freeway, but increased when stopped at a traffic signal (Lerner et al., 2005). The data presented in this chapter suggested a similar pattern. The freedom of the open-ended questions in the semi-structured interviews allowed participants to infer their own interpretations of the roadway and their motivations for engaging with the technological tasks. Drivers reported a total of 462 references to the 'Infrastructure' themes across the 30 interviews. There was a clear distinction, across tasks, between the likelihood of engaging while driving through a junction versus when stopped at a junction (with the exception to listening to music, which is a continuous task that did not alter across the junction). Interviewees were provided with images of the road types, the one corresponding to 'stopped' at a junction displayed a traffic light which led participants to discuss the road event as a definite stop, rather than waiting for a gap in the traffic at an intersection. This may suggest the high ratings of engagement. Despite UK laws not distinguishing between the use of phones when the vehicle is stopping in traffic versus driving, participants of all ages were more likely to state they would interact with their phone when stopped at lights or in traffic, apart from the older females who have been found the least likely demographic to use their phone while driving (Pöysti et al., 2005). Comments such as *"Just because if I'm stopped at a junction I'd have my hand-brake on and my car's stationary so I won't personally cause a crash"* (participant 7) suggest that drivers do not view the risks to be as great when the car is stationary. Many stated that if they were stopped that would be when they would 'have a go' at doing the task.

The likelihood ratings are varied across the other road types with no clear trends, although there may be some notion that drivers are slightly more likely to engage with tasks on the motorway, but this is by no means conclusive and there are inconsistences between tasks and participants. The inconsistencies may be due to the various other factors not accounted for, such as varying road conditions often found across road types (Horberry et al., 2006), drivers' perceived familiarity with the road (Charlton & Starkey, 2011), the journeys they are making when accessing different roads (e.g. longer journeys using motorways) and the confidence of drivers when driving on different road types. Distinctions between the road types on drivers' situational awareness constructs were identified by Walker et al. (2013) when drivers were actually driving the roads, rather than talking hypothetically. It would therefore be useful to explore how the drivers' interpretation of how likely they would be to use the technology when they are actually driving on the road would differ. This would provide more contextual information surrounding their reported likelihood of engaging.

7.3.3.7 Legislation

The 'Illegality' theme relates to road safety laws, policy and regulation and is included under the 'Infrastructure' main theme, in accordance with other taxonomies (e.g. Stanton & Salmon, 2009; Salmon et al., 2010). The distractive effects of mobile phones have led to the decision to ban their use by drivers across many countries, yet hands-free alternatives are permitted as well as other technologies such as satnavs and in-built IVIS features. Chapter 5 suggested that the prohibition of mobile phones may infer that other technologies are comparatively safer to use which led to the efficacy of legislation as a distraction mitigation technique to be questioned. There were only 19 references coded to the 'illegality' theme, this is small in contrast to 251 references to 'interaction' and 95 to 'task context'. Hence, drivers did not seem overly influenced by legislation when reasoning why they be more or less likely to engage with technological tasks.

7.4 IMPLICATIONS

This chapter has presented findings from a semi-structured interview study that assessed drivers from different age and gender categories on their self-reported likelihood of engaging with technological devices, and their reasoning why. With technology becoming increasingly accessible to drivers, they are given more choice in when they can engage with technological tasks. These include tasks that do not bare any enhancement to the safety of the driving task and may instead be detrimental to it. The findings presented here suggest that drivers do report that they are likely to engage in illegal and distracting activities while driving and the reasons for this are attributed to a multitude of factors. The influence of age and gender have found similar effects as those previously stated in the literature (e.g. Lamble et al., 2002; Pöysti et al., 2005; McEvoy et al., 2006; Pope et al., 2017). The qualitative analysis has, however, revealed how drivers discuss the factors that influence their decision to engage with technological tasks. While some drivers, particularly those in the older age category, are able to restrict their use of technological devices while driving, others state that they are quite likely to engage with tasks on their satnav, IVIS, hands-free or handheld mobile phone that may result in them becoming distracted from the driving task. The findings from these discussions has implications for distraction mitigation strategies including those that target actors who can control the drivers' interaction with the devices (e.g. manufacturers, companies, regulators), rather than letting the end user decide (Parnell et al., 2017a).

Satnavs are now commonly used by drivers to navigate, with the egocentric display allowing the driver to easily find their destination (Speake, 2015). Most drivers

Technologies People Use When Driving and Why 141

within this study said they would only use the satnav if they did not know where they were going, although some stated they would use it even if they did know where they were going to update them on the traffic and their time of arrival. This suggests that the utility of the device has extended past the navigation function, and drivers are displaying additional navigation information even though they do not require it. The development of navigation apps on mobile phones has also made them more accessible and cheaper, although it was found that drivers tended to state that they were more likely to enter a destination on the satnav device than their mobile phone. The lack of references to the illegality theme when discussing satnav use suggests that the participants were not aware of any regulations that were imposed on it, but a number of references were made to the features of the task, with some commenting on the complexity and multiple inputs required to enter a destination.

Previous findings suggest that tasks within the vehicle should take no longer than 12 s to complete in total, comprised of segments no longer than 2 s (NHTSA, 2013). Yet, evidence that the task of entering a destination while driving takes longer than 12 s suggests that the task should be disabled while the vehicle is in motion (NHTSA, 2013; Harvey & Stanton, 2013). While some participants stated that they would not change a destination while driving, preferring to stop or enter it before they started driving, other drivers said that they would perform this task while driving and have done so in the past. Manufacturers often display information on the use of the device to the driver when the device is turned on, to warn the user of the potential for distraction and reducing the manufacturer's liability for misuse. Yet, a more proactive approach would be to disable tasks that are known to be distracting. While some manufacturers do disable some features, including address entry, more regulations across the industry are required to determine which tasks should be prevented from use in motion, or even across certain road environments. Car manufacturers that take a more restrictive approach to managing the drivers' interaction with devices do seek to disable functionalities, such as the Japanese market (NHTSA, 2013). Yet, the European market place more responsibility with the driver and therefore allow them more choice in the interaction with devices. While these separate cultures are considered within the NHTSA (2013) guidelines, the drivers' responses in this study suggest that they are likely to engage with the devices if they have the opportunity to do so and that enabled functionality may lead drivers to believe that the device is safe to use in motion.

Drivers stated that they would be highly likely to engage with the range of facilities within the IVIS. A pertinent comment was made by one participant that they felt it was ok to engage with these tasks as, by placing the technologies within the vehicle, the manufacturer was saying it was ok for them to use them while driving. This further highlights the role that vehicle manufacturers have in facilitating, and prohibiting, distractions within the vehicle. Furthermore, regulations and guidelines on the design and functionality of in-built information systems must ensure manufacturers are only able to place safe and easy-to-use displays within the vehicle that do not distract the driver. Speech recognition technology offers an alternative way for the driver to interact with in-vehicle technology, using voice commands that do require them to take their hands off the wheel or their eyes off the road. In contrast to manual button press interactions, there was more comments directed towards the desirability and ability to engage with voice command systems. There was a clear distinction in the use of voice command systems between the age categories, which appeared to stem from the drivers' trust and views on the reliability of the system. Younger drivers were keener to have a go at using the system, with ratings as high as the other IVIS tasks. Conversely, it was evident that the middle and older age groups were not confident in the system which limited their use of it. It is clear that the implementation and reliability of the voice command technology has some way to go before its benefits are realised by drivers, particularly older drivers who may be more used to traditional ways of interacting.

In line with other conclusions in the literature, younger drivers (e.g. Lamble et al., 2002; Pöysti et al., 2005; McEvoy et al.., 2006; Young & Lenné, 2010), and the middle age male group (Engelberg et al., 2015; Pope et al., 2017) reported themselves to be more likely to engage with technologies while driving than the older age group. The older age group were found to differ on their attitudes towards technological tasks and their physical ability to interact with them while driving, further supporting previous findings in the literature (Strayer & Drews, 2004; Walsh et al., 2008; Pope et al., 2017). Drivers stated that they would, generally, be more likely to engage with tasks on the hands-free phone than on the handheld phone. The legislation in the UK that permits the use of hands-free phones but prohibits handheld mobile phone use may be one reason for this. The bans on mobile phone use propelled the industry to respond with advanced hands-free communication devices that allow drivers to be contactable while driving. Yet, research has suggested that hands-free communication is no safer than handheld mobile phone use (Strayer & Johnston, 2001; Horrey & Wickens, 2006). While the hands-free aspect limits the physical aspects of the distraction, the cognitive component of talking on the phone still has a negative impact on the driving task when using a hands-free device (Tornros & Bolling, 2005; Treffner & Barrett, 2004). Despite this evidence, hands-free devices are still permitted and are even readily built-in to modern vehicles by manufacturers that are then actively used by the driver who may not be aware of the increased risk talking on the hands-free phone may cause.

Although drivers do state that they would be more likely to use a hands-free phone than the handheld phone, there was still a concerning number of participants, even in the small sample of interview participants, who claimed they would be likely to engage with their handheld mobile phone under some circumstances while driving. The claims by these drivers provide evidence for the argument that the use of penalties and fines enforcing the ban on mobile phones in the UK are inadequate. Many countries have increased penalties for young or novice drivers that aim to target this high-risk population, yet the results of this study, and others (e.g. Young & Salmon, 2015; Parnell et al., 2017a), suggest that the law alone is not enough. Therefore, it is recommended that alternative, systemic, measures for targeting the use of technologies by younger drivers be explored further. For example young drivers engaged in a lot of discussion on the features of the tasks, their interaction with them and the ease of doing so, in contrast to older drivers. Therefore, this aspect of technology interaction should be investigated for practical strategies to prevent distraction.

Reading a text message was the task that drivers from all age groups rated as more likely than other mobile phone tasks. It is argued here that the reactional nature of this task is what affords this higher rating, with many drivers commenting that they are often alerted to the text by a tone or it flashing up on their phone, which catches their attention and encourages them to then read the text or see who it is from. This therefore questions the responsibility of phone manufacturers who permit such interactions with drivers. The technology is available to impose phone applications that freeze phone interactions while driving but they are not in widespread use. Phone manufacturers currently have no obligation to cater for phone functionality while driving; advocates of a more systemic approach to safety state that this should change (e.g. Young & Salmon, 2015). Furthermore, as drivers are increasingly using their mobile phones for multiple functionalities while driving, such as music players and navigational aids, phone manufacturers need to be aware of how their devices are influencing the driving task and their potential to cause distraction-related accidents. One step towards this is the latest software update for iphones (ios 11, released September 2017) by Apple that incorporates a 'do not disturb while driving mode' that will sense when the phone is in a moving vehicle and will, once prompted by the user, turn off notifications, and prevent the screen from lighting up (Apple, 2017). The feature can also be set up to auto reply to contacts to inform them the person they are trying to contact is driving. This is a feature that other device manufacturers have tried to develop. This is a large step by a leading phone manufacture in realising their responsibility for distracted drivers. Future research is needed to assess the efficiency of this tool as well as its uptake and integration with general mobile phone use. Such methods can be subject to user adaption to overcome boundaries. Furthermore, the current version of the application can be disabled readily by the user or overridden while in motion.

The possibility of phone multifunctionality when driving also influences where drivers choose to position their phone. Many of those who stated they use their phone for navigational purposes fix their phone to the vehicle in a position that is easy to see. Aftermarket manufacturers have developed fixings for this purpose in accordance with the UK law that bans any handheld interactions with the phone but does not prohibit the placement of the phone in the drivers' line of sight, as long as it is fixed to the vehicle. Drivers are therefore more likely to be alerted to text messages and incoming phone calls, increasing temptation to engage with them. This may be where 'do not disturb while driving' features may be particularly important. This further highlights the workarounds that occur when legislation specifically bans certain tasks that can lead to further issues rather than resolving existing ones (Parnell et al., 2017a). Distraction mitigation strategies need to be aware of these issues and the desire for multifunctionalities of smartphones, such as navigation assistance which is not specifically banned for use while driving. With technology becoming increasingly accessible to drivers, they are given more choices of when they can engage with technological tasks that do not bare any enhancement to the safety of the driving task and may instead be detrimental to it. This enforces the notion that other actors who do have control over device use (e.g. manufacturers, companies, regulators) need to be more pro-active in preventing such high-risk decisions from being left to the control of the end user.

Across the majority of the tasks presented, drivers of all age groups tended to report that they are more likely to engage with technologies when stopped at a junction, for example at a red light, and less likely when they are driving through a junction. This suggests that drivers do adjust their decision to engage in line with the demands of the environment. Participants generally perceived the consequences of their actions to be reduced when stopped at a junction, as they were not perform-ing the act of 'driving'. Yet, engaging with distracting tasks is still prohibited when stopped in traffic or at a red light as drivers need to be alert and aware of other traffic at all times. Statistics from a naturalistic study support this finding with double the percentage of drivers observed to be using their phones when stopped in traffic than when the vehicle was in motion (Department for Transport, 2017). If the highway code and road safety laws, that aim to regulate the use of portable handheld devices while driving, are to be optimised then the distinction that drivers currently make between being in-motion and stationary needs to be questioned. Alternatively, approaches such as those taken in Sweden could be adopted, which permit the use of technologies, including mobile phones, by drivers as long as they do not engage in activity that may be 'detrimental' to their driving. This encourages the driver to take responsibility for when they choose to engage with technologies and reduces any assumptions that some technologies that are not banned, or built into the vehi-cle, are safer than those that are banned. If drivers are able to strategically adapt their driving engagement with technological devices to facilitate their safe engage-ment while driving, then, given the choice, drivers may be able to engage with their devices at optimal points without adverse safety implications. This is, however, con-troversial to the current views across most countries that aim to prevent the driver from engaging in any distracting activity. It would require substantial research and evidence to assess the advantages to its implementation across other national road safety programs.

7.5 EVALUATION AND FUTURE WORK

The qualitative interpretation of the quantitative ratings has allowed the explora-tion of the prevalence of key concepts that were discussed by drivers of different ages and genders towards different technology types. Within the semi-structured interview study, a trade-off was made between gathering in-depth qualitative data for an in-depth analysis and sample size. The limited sample means that caution should be taken in the interpretation and generalisations made between the data and the wider driving population, particularly when looking at gender effects. Future research should seek to explore the qualitative themes generated to a larger population of drivers. Additionally, the findings are limited to some degree by the self-report methodology, future research should seek to research these concepts within realistic driving conditions. Ethical considerations surrounding driver distraction research can prevent researchers from capturing drivers' physical

Technologies People Use When Driving and Why **145**

engagement with distractions under real-world conditions (Young et al., 2008). Yet, asking drivers to verbalise their willingness to engage with distractions in simulated and naturalistic driving environments is an interesting avenue for future research that is explored in the next chapter.

7.6 CONCLUSION

An understanding of why distraction-related accidents occur has been under-researched. This chapter and Chapter 6 have demonstrated the utility of semi-structured interviewers in gaining insights into the drivers' perceptions of why they may be likely to engage with technological devices while driving. The previous chapter developed the inductive thematic framework to reveal the key themes that drivers state have an influence on their intention to engage with distractive technologies. This chapter has shown the utility of applying this thematic framework, alongside quantitative ratings that are more commonly used in the field with the application of surveys (e.g. Walsh et al., 2008; Young & Lenne, 2010; Zhou et al., 2012). The novel application of semi-structured interviews has generated both quantitative data to understand *what* tasks drivers were likely to engage with and qualitative data to propose the reasons *why*. Correlations of the quantitative data to a larger sample of drivers suggests that the findings are representative of drivers outside of those sampled. The use of the NVivo 11 software to code the interview transcripts to the thematic framework developed in Chapter 6 permitted further quantitative analysis in the form of a matrix query to identify what key concepts were discussed by the participants in relation to the different technologies. The findings have shown support for previous age and gender differences in what tasks are engaged with, as well as suggesting how these differences arise due to the themes that drivers from different demographic groups discuss when stating their reasoning why. Although, due to a trade-off between in-depth qualitative analysis and sample size, the interview sample size is small in contrast to that which can be recruited with surveys, therefore data should be interpreted with caution.

Areas for concern include points where drivers think it is acceptable to engage with technologies, such as when stopped at a traffic light. The location of portable devices was also highlighted as highly variable across drivers. This has been under reported in the past, yet, results presented here found it to influence how likely drivers were to engage with tasks on the device, especially reactional tasks like reading a text message. The use of mobile phones is also heavily interlinked with their multi-functionality and the high level of young participants who report using their phone as a satnav, which increases their interaction with the device while driving. To prevent distraction occurring from these sources, it is suggested that manufacturers of portable devices and in-built technologies, including satnavs, information systems and hands-free devices, should take more responsibility over the use of their devices at in-appropriate times, such as when driving. Moreover, it was also highlighted that the high use of IVIS tasks may relate to the perceived safety of completing these tasks as they are designed for use while driving. Therefore, manufacturers should account for this when designing driver interfaces. A more proactive approach is needed that prevents high-risk decisions from being left to the control of the end user.

The participant discussions in the interview study suggest that they are aware of the demands and consequences of their interactions and therefore may be able to strategically engage with technologies in line with the demands of the environment. The semi-structured interviews are, however, limited by their reference to behaviours outside of the context from which they arise, with drivers talking hypothetically about their behaviours. Further work is required to explore the application of the thematic framework initially identified in the interview setting to the driving setting to assess its validity and the rated likelihood of engaging with the technological tasks. This is the focus of the next chapter.

8 Good Intentions? Willingness to Engage with Technology on the Road and in a Driving Simulator

8.1 INTRODUCTION

This book has suggested that it is useful to understand why driver distractions are occurring in order to understand the root cause of the issue, rather than just focusing on the consequences once they have occurred. In an analysis of crash reports, Beanland et al. (2013) found 70% of distraction-related crashes to be voluntarily engaged by the driver. This suggests that distractions occur as a result of the individual making the considered decision to engage with secondary tasks, despite being aware of the risks. Although, it has also been suggested that drivers are skilled at self-regulating when to engage with secondary tasks if they can predict the upcoming demand (Tivesten & Dozza, 2015). The previous two chapters have gained an understanding of the reasons that drivers give for their engagement with technologies in a self-report, interview-based, study. While this may support the suggestion that drivers volunteer themselves to become distracted, it has also demonstrated the relevance of other elements in the road transport system that influence the drivers' behaviour, such as the context, infrastructure and the task itself. Yet, to review the ecological validity of the decisions that drivers make to become distracted, the area of naturalistic decision making (NDM) is of interest.

8.1.1 NATURALISTIC DECISION MAKING

Traditional decision-making theories suggest that an individual makes their choice from a set of alternatives using optimisation strategies generated from identifying multiple options, probabilities and utility estimates (e.g. Edwards, 1954). More recent understanding suggests that decisions are actually based on intuition (Phillips et al., 2004), experience (Klein et al., 1986), biases and heuristics (Kahneman et al., 1982). These ideas have emerged from reviewing decisions within their naturalistic environments through NDM methods that seek to understand cognitive processing in relation to the demands and complexities of the situation that they are occurring in.

148   Driver Distraction

The process of satisficing which relates to finding the first 'good enough' option rather than the best possible option (Simon, 1957) has been realised when assessing decision makers in their naturalistic environment (Klein, 1989). This phenomenon has also been linked to driver distraction in the minimum required attention theory (Kircher & Ahlstrom, 2016) which incorporates the notion that driving is a satisficing and self-paced task (e.g. Boer & Hoedemaeker, 1998; Summala, 2007). Kircher and Ahlstrom (2016) deem drivers to be attentive when they are collecting 'enough' information from the road environment as is needed relative to the current demands, allowing the driver to adapt their behaviour and engagement with secondary tasks in line with the demands of the situation. This view of distraction suggests it to be heavily related to the situation in which it arises, which complements the findings that drivers adapt their behaviour in line with the situational demands (e.g. Brookhuis et al., 1991; Haigney et al., 2000; Rakauskas et al., 2004; Parnell et al., 2016). When assessing distraction, it is therefore important to account for the situational demands and the context surrounding the behaviour (Sharples et al., 2016).

An assessment of the drivers' engagement with secondary tasks in relation to their situational awareness suggested that drivers do strategically plan their decision to engage with a secondary task (Schömig & Metz, 2013). The planning level of engagement was one of three levels of adapted behaviour that influence the engagement of secondary tasks alongside the primary driving task. The decision level occurs when drivers decide to go ahead with interactions with secondary tasks in line with the demands of the driving situation. Control level adaption occurs once the secondary task has been engaged with and drivers adapt their driving behaviour, for example by slowing down (Hockey, 1997), or adapting their secondary tasks interactions, for example interrupted glances (Harvey & Stanton, 2013), in order to integrate the two tasks. Thus, tightly linked to the drivers' decision to engage and subsequent completion of the task is the adaption of behaviour in line with the demands of the situation and the task (Schömig & Metz, 2013). This is particularly true at the control level, with evidence to suggest that drivers adapt their speed (Cnossen et al., 2004), following distance (Alm & Nilsson, 1995) and glance behaviour (Metz et al., 2011) in line with the demands of the environment. These levels have also been discussed by Michon (1985) as the strategic (planning), tactical (decision) and operational (control) levels of problem-solving engaged by the driver.

'Adapt to demands' emerged as a key mechanism of distraction in the PARRC model (Chapter 4; Parnell et al., 2016), which captured the prominence of the drivers' adaption of their behaviour to manage secondary task goals alongside the primary driving goal. Furthermore, the discussions obtained in the semi-structured interview study in Chapters 6 and 7 of this book suggested that drivers adapt their engagement with technological devices at the strategic level by planning their engagement for specific points. For example they stated that they would be more likely to engage with devices when stopped at lights than when driving through a junction. Evidence for adaption at other levels was also evident across other themes of the framework that included the reports that drivers perform the task without taking their eyes off the road under the 'interaction' subtheme (decision level) or slow down to engage with the device under the 'task-road relationship' subtheme (control level). Yet, the notion that drivers can strategically plan when to engage with distracting tasks is

Willingness to Engage with Technology

pertinent as it suggests that they can self-regulate their engagement in line with the demands of the driving task. Yet, findings on the strategic adaption of secondary task engagement have been mixed in the literature. This may, however, relate to the setting in which the behaviour is explored (Horrey & Lesch, 2009); this is discussed in the next section.

8.1.2 Experimental Setting

The challenges faced in the assessment of driver distraction are discussed in Chapters 2 and 3, where it is highlighted that capturing driver behaviour in its naturalistic environment is particularly difficult (Carsten et al., 2013). A trade-off between validity and control is central to the selection of methodologies and the experimental setting within the domain (Burnett et al., 2004). The road safety risks that distracted driving poses has meant that many studies that aim to observe the consequences of distraction have utilised driving simulators (e.g. Summala et al., 1998; Tsimhoni et al., 2004; Reimer, 2009; Young & Lenné, 2017) or test-track facilities (e.g. Ranney et al., 2005; Horrey & Lesch, 2009). These provide a controlled and safe environment to trial new technologies and measure driving performance. Technological advancements have vastly improved driving simulation facilities, improving the fidelity and validity of experimental studies (Young & Lenné, 2017). This is to the advantage of distraction research that is often unsafe and unethical to perform on the road. Such facilities have enabled an understanding of the adverse consequences of secondary task engagement when the onset of the task is predetermined and controlled within the experiment (e.g. Lansdown et al., 2004). Yet, if it is understood that distractions are largely voluntarily engaged by drivers (Beanland et al., 2013), and engagement is largely determined by situational factors, then the pre-determined scenarios used in research studies cannot be readily generalised to realistic scenarios where drivers choose to engage with secondary tasks. They fail to account for any strategic engagement with the secondary task which may offset any adverse effects. For example, Lee and Strayer (2004) highlight the overestimation of risk attributed to older drivers engaging with handheld phones from findings that force drivers to engage with the technology, when older drivers may not actually choose to engage with the phone at all when driving in their everyday lives.

Furthermore, the level of reliability, validity and fidelity established within driving simulations needs to be reviewed when assessing findings and their contribution to understanding (Young & Lenné, 2017). Advancements in technology have led to the development of simulators with enhanced fidelity. Driving cabs have been found to increase the drivers' motivation and reduce the potential for simulator sickness in contrast to the desktop-based simulators (Burnett et al., 2007). The validity of driving simulators is more complex to assess, and there are multiple different forms of validity to determine, including physical, behavioural, relative and absolute validity (Young & Lenné, 2017). While physical validity involves the representation of the simulator components to the real-world driving task, behavioural validity refers to the extent to which the driving simulation can emulate real-world driving behaviours. Absolute validity requires the matching of numerical values between the road and the simulator (Godley et al., 2002), but it is hard to achieve (Godley et al., 2002;

Underwood et al., 2011). Whereas relative validity asserts that differences between the road and simulator environments are in the same direction (Godley et al., 2002). Numerical measurements such as vehicle dynamics like speed and lane keeping are stated by Underwood et al. (2011) to be important but not sufficient, as behavioural patterns are important to consider.

The decision to engage with distractions has been explored under different experimental conditions with differing findings. The semi-structured interview study presented in Chapters 6 and 7 asked drivers to state their intention to engage with technological tasks with only images and descriptions as cues to the road environment. They were asked to detail their likelihood of engaging with technological tasks across different road environments that have been found to influence the drivers' awareness of their surroundings (Walker et al., 2013). This aimed to explore how drivers would discuss, hypothetically, what would influence the likelihood of engaging with different technological tasks. The inductively generated thematic analysis highlighted the complexity of factors that influence the drivers' decision to engage, as interpreted from the drivers' own perspective (Parnell et al., 2018a). Yet, the interviews were only able to capture the drivers' hypothetical reasoning for engaging with technologies while driving. They did not capture the context surrounding the interaction and how this may impact on intention. Holtzblatt and Jones (1993) highlight the importance of describing behaviour in the context within which it is produced. The knowledge that has been gained through exploring decisions in their natural environment, as opposed to in highly controlled laboratories, has developed understanding across many domains (Klein, 2008). A description of behaviour outside of its context is a description that is based on ways that it could or should be performed, but fails to capture the processes involved in the behaviour (Holtzblatt & Jones, 1993).

Schömig and Metz (2013) conducted a study in a driving simulator setting to assess the drivers' decision to engage with, and consequently complete, a secondary task while driving. The prompted decision was compared in critical and non-critical situations to reveal that drivers were less likely to decide to engage in critical situations. Thus, suggesting their strategic adaption of the secondary task in line with the demands of the driving environment. They also found that engagement was informed by the drivers' attitudes and perceptions about the task. They proposed that drivers use global strategies to determine when engaging with distractions is possible (Schömig & Metz, 2013).

Horrey and Lesch (2009) explored the drivers' decision to engage with driver-initiated secondary tasks while driving on a test track. They allowed drivers to freely choose when to engage with the distractions to determine if their choice was strategically adapted to the demands of the route; this increased the validity of the process in contrast to Schömig and Metz (2013), who predefined the decision stage. Yet, in contrast to other findings (Lerner, 2005; Schömig & Metz, 2013), they found no evidence to suggest drivers did strategically adapt their driving performance. Rather, Horrey and Lesch (2009) claim drivers chose to engage with the task independently of the road environment or task type and then breakdown the task into chunks to integrate the secondary task and primary task. This supported previous findings that drivers adapt their interaction at the control level (Haigney et al., 2000; Strayer et al., 2003)

Willingness to Engage with Technology

but not the strategic level. Yet, there were some aspects of the study that requires further assessment, for example the use of a test-track environment and lack of real-world motivations were cited as possible reasons for the lack of evidence for strategic planning (Horrey & Lesch, 2009). Although the driving task is more realistic when driving on a test track in contrast to a driving simulator, drivers are still not exposed to 'true risk' of driving on the roads or interacting with other road users who are likely to influence their decision to engage.

Some naturalistic driving studies have sought to explore the drivers' decision to engage with technology while driving by video recording the drivers' behaviour over periods of time (Metz et al., 2015; Tivesten & Dozza, 2015). They have demonstrated that the drivers' engagement with distractions is self-regulated and that drivers are able to strategically engage with tasks in line with predictable demands in the environment such as sharp bends, but that these strategies are ineffective when faced with unpredictable events such as a lead vehicle braking (Tivesten & Dozza, 2015). Such naturalistic driving studies are, however, limited by their inability to understand *why* drivers engage with distractions as they only capture *what* drivers do (Tivesten & Dozza, 2015). Further research is required to assess the drivers' decision to engage with distracting tasks within naturalistic conditions, as well as how this may vary to the use of driving simulators, which are a useful and commonplace instrument in the assessment of driver distraction.

This chapter presents a study that aimed to assess the drivers' decision to engage with technological devices while driving in a simulator and on the road in an Instrumented Vehicle (IV). Scenarios where drivers may need to interact with technology were posed to participants while they drove along a pre-determined route. The drivers' decision was assessed at set points along the route through verbal protocol responses to the scenarios posed. The verbal protocols informed on the drivers' intention and, importantly, their decision-making process. The drivers' decision to engage and the contents of the verbal protocols were compared across the simulator and road condition. The potential for this method to be used to assess distraction without causing adverse effects to driving performance was also explored.

8.2 METHOD

8.2.1 PARTICIPANTS

Participants were recruited from the original sample of participants in the interview study. Twelve participants (six males and six females) volunteered to take part, including two males and two females from each of the three age categories (young, middle and older) detailed in Chapter 7. The demographic information of these participants is presented in Table 8.1. All participants were required to hold a full driving licence, with no more than three penalty points, and be over the age of 25. This was required for the university insurance policy for the instrumented vehicle (IV). Participants were also required to be frequent drivers, driving a minimum of twice weekly, and live local to Southampton in order to be familiar with the roads of the test route. Ethical approval was granted from the University's Ethical Research Governance Office (ERGO 26046).

TABLE 8.1
Demographics Table

Demographics	Mean	SD	Range
Age (years)	39.75	11.8	26–58
Annual mileage (miles)	10,733	4,530	6,000–21,000
Age of main vehicle (years)	5.8	3.1	1–12
No. of years since passing driving test (years)	18.5	12.2	7–40
No. of hours spent driving weekly (h)	8.25	5.7	4–25

8.2.2 EXPERIMENTAL DESIGN

A repeated measures design was used with two driving conditions (simulator and IV). The drivers' decision to engage with a task in four different scenarios (read a text, change a destination, make a phone call, change a song) across three road types (motorway, A road and roundabouts) was assessed. The decision to engage (yes or no) was measured through verbal protocols that included the drivers' stated intention to engage with the task and their reasoning why. Voice recordings were transcribed and coded. Driving speed was also measured in both conditions. All scenarios were posed to participants across all three road types, with measures taken to ensure this occurred at the same points along the route when they drove on the road and in the simulator for all participants. The order of the simulator and road conditions were counterbalanced, as were the order of the scenarios across the road types.

8.2.3 EQUIPMENT

8.2.3.1 Vehicles

The vehicle used in the on-road trial was a Fiat Stilo (Figure 8.1a), an IV owned by the University of Southampton. It was a right-hand drive with automatic transmission. The Southampton University Driving Simulator (SUDS) was used for the simulator condition (Figure 8.1b). SUDS is a fixed base, right-hand-drive Land Rover Discovery with automatic transmission. The simulator environment was created

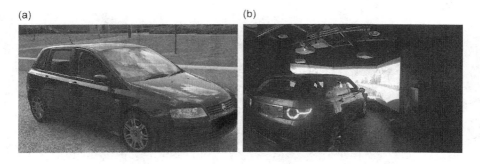

FIGURE 8.1 Image of the vehicle used in the road condition (a) and simulator condition (b). Credit to Fiat S.P.A and Land Rover.

Willingness to Engage with Technology 153

with STISIM M500W wide field of view and was projected onto three screens in front of the vehicle and one behind, to give the rear view. The simulator has a 135° field of view.

8.2.4 PROCEDURE

The study took approximately 2 h to complete and was comprised of multiple stages that are summarised in Table 8.2. The study was conducted between 10.30 am and 3 pm on weekdays only, to ensure that the conditions on the road avoided periods of heavy traffic (i.e. rush hour). The study was also only conducted in dry weather conditions. Participants were given an information sheet on arrival before being asked to give signed consent to participate. They were guided through the stages in Table 8.2 by the researcher until the end of the experiment, where they were reimbursed £10 for their time.

8.2.4.1 Verbal Protocol Methodology

Verbal protocol is a method that has become increasingly prevalent in driving behaviour studies (e.g. Young et al., 2013a; Banks et al., 2014; Salmon, Walker et al., 2017). The method aims to capture the contents of an individuals' working

---

TABLE 8.2
Breakdown on the Study Procedure Into its Comprising Stages

Study Stages		Approx
START	Introductory briefing	5 min
	Training in verbal protocol (including a video session)	10 min
	Pre-trail interview to discuss scenarios	10 min
	Drive 1: Simulator/Road (including practise session)	40 min
	NASA TLX questionnaire	5 min
	Drive 2: Simulator/Road (including practise session)	40 min
	NASA TLX questionnaire	10 min
	Post-trails interviews to debrief	10 min
	Participants were paid £10 at the end of the experiment	Total: Approx. 2 hours

memory through their verbal reports in relation to the context and decisions that they are making (Ericsson & Simon, 1993). It essentially requires participants to 'think aloud' and generate a trace of their cognitive processing and the sequence of its unfolding from the point of view of the individual. The cognitive events can be traced from the point at which new information is introduced to a setting, and the decision outcome that is then achieved. In this study, participants were required to produce verbal protocols of their decision-making process when posed with the task scenarios. Importantly, drivers were not required to actually complete the tasks, they were just asked to verbalise their intention to engage, given the current status of the road environment. This aimed to capture the participants' perceptions of the environment and how it would influence their strategies to engage with the technologies while driving.

Drivers were given training in how to perform the verbal protocol across three stages. First, they were given a written overview of the verbal protocol procedure and what it was aiming to uncover, that is thinking aloud to produce a continuous stream of the drivers' working memory (Ericsson & Simon, 1993). The procedure, and what was expected of them during the study, was further explained by the researcher. They were then shown a video of verbal protocol experts participating in the road condition of the study, responding to the same scenarios in the same conditions that the participants would be asked. Participants were encouraged to ask questions at any point if they were unsure. The final stage allowed the participants to practise producing their verbal reports while driving during the practise phase of the route (see Section 8.2.4.3 for further details of the route). Both the simulator and road condition had a practise section that not only allowed the participant to acclimatise to the vehicle but also to then practise responding to the scenarios that the researcher posed to them while driving. This practise section lasted approximately 10 min. The researcher gave feedback to the participant during this practise stage until they were confident that they were proficient. The researcher also gave the participants prompts in the experimental testing phase to make sure they elaborated on their verbal reports. Participants were required to wear a head-mounted microphone that enabled high-quality voice recordings that could be picked up over the noise of the vehicle while driving.

8.2.4.2 Task Scenarios

To provide drivers with a consistent motivation for engaging with technological devices, predetermined scenarios that would require technological engagement were developed. The scenarios are presented in Table 8.3. The tasks were selected from the list of tasks given to drivers in the interview study as ones that generated interesting discussions with the drivers as well as being of interest to the literature. Scenarios were developed based on what the participants in the interview study stated to motivate them to engage with the technological device (see Parnell et al., 2018b).

Read a text: The findings presented in Chapter 7 showed that reading a text was the task drivers rated to be the most likely to engage with on a handheld mobile phone. Many drivers stated they would often glace at who the text was from and then make the decision on whether they would go on to read the text if they felt it was important enough. It was suggested in Chapter 6, when applying the thematic framework to the PARRC model of distraction, that this process may reflect an association

Willingness to Engage with Technology

TABLE 8.3

Scenarios Relating to Different Technology Use

Task	Scenario
Read a text	You receive a text message which sets off the alert tone and causes the phone to light up. Being aware that you have received a text message. Would you read the text at this moment in time?
Change a destination	You are driving on a long journey to a destination you are not familiar with and therefore are using a satnav to direct you. Along the way, you realise that you need to stop off at an alternative destination that is not currently programmed into your satnav. Would you enter the new destination into the satnav?
Make a phone call	You are driving to a meeting but you are running very late due to heavy congestion and you will not be able to arrive at the time you prearranged. Would you call the person you are due to meet to let them know you are running late?
Change a song	You are driving while listening to the radio when a song comes on the radio that you really do not enjoy listening to. Would you change the radio station to find one that is playing music that you do like?

between the 'resource constraints' and 'goal priority' factors that was not evident in the literature. This therefore required further assessment.

Change a destination: This task was chosen as it reflects a task that is not widely acknowledged to be illegal but, as demonstrated in the case study of Victoria McClure (Chapter 4; Parnell et al., 2016), can have an adverse effect on the drivers' attention towards the road. Likelihood ratings in the interview study were more varied in relation to the enter-a-destination task, with many discussions on engaging with the satnav relating to the road infrastructure. Younger drivers stated that they would be more likely to use their phone to navigate than a satnav, but that they would be more likely to change a destination on an actual satnav device than their phone. This may be due to the law which bans the use of mobile phone interactions while driving, yet as younger drivers were also likely to engage with their handheld mobile phone, this may not be the case. Further research on the use of navigation systems was required.

Making a phone call: This task was chosen, again due to the controversy of mobile phone use while driving and the legislation against the handheld phone use. However, the method of making the phone conversation was not predetermined as the use of hands-free phone was also explored to see how drivers may approach the task while driving. Hands-free phones are road legal, although they have not been found to be any safer than handheld alternatives (Alm & Nilsson, 1994; Horrey & Wickens, 2006). Yet, as suggested in Chapter 5, drivers may deem them to be safer. The ways that drivers may approach this scenario while driving were explored.

Change a song: In Chapter 7, drivers stated that they would be the most likely to engage with the tasks on the IVIS as they perceived these to be easy to complete and, as one participant stated, put there by manufacturers to be safe to use while driving.

Participants were first presented with the scenarios in Table 8.3 during the introductory briefing. They were interviewed by the researcher on how they thought they would typically respond in these scenarios. This included asking if they had ever come across the scenario before, if they have the technologies mentioned available to them in their own vehicle, and if so, how do they typically interact with them. The interview was recorded on an Olympus digital voice recorder. These interviews enabled the researcher to establish the drivers' initial response to the scenarios and also how they would typically go about making the interactions with the technology, for example do they use voice-assisted commands, steering wheel buttons or hands-free phones. These could then be referenced by the researcher during the trials for clarity if necessary. It was also useful to establish the different ways in which the four common scenarios could be responded to.

8.2.4.3 Route

The same pre-determined route was driven in both the simulator and the road condition. The route was chosen as it started and ended at the University and covered the different road types of interest (A road, motorway and roundabouts). It was 13.2 miles long, taking approximately 40 min to drive (including the practise segment). The simulated route was designed to replicate the real world as much as possible using the STISIM software. This included road curvature, traffic density and the development of salient localised objects with the creation of a custom library of models that aimed to replicate key objects and landmarks from the real-world route. For further information on this route, as well as the pilot study, see Allison et al. (2017).

Figure 8.2 shows a map of the route that the participants were given during the introductory briefing. Participants were encouraged to familiarise themselves with this route, although they did not need to memorise it as the researcher sat in the passenger seat of the car in both the simulator and road condition to provide directions. The blue line shows the test route, with the arrow indicating the practise segment. The structure of the study required participants to state their intention to respond to the technologies in each of the four scenarios on each of the three road types: motorway, A road and roundabouts. The points along the route where the scenarios were posed are shown in blue (motorway), red (roundabouts) and green (A road/urban road). This required participants to produce 12 verbal protocol reports. The key on Figure 8.2 shows where the researcher asked the participants to state their intention to engage with the technology in the scenarios. The researcher conducted all the trials for consistency to ensure that they were asked at the same points across all participants. Verbal reports lasted approximately 60 s. The order in which the scenarios were asked was randomised between participants as was the order of the simulator and road trials. Yet, within participants, the scenario order was the same in the road and simulator condition. The researcher made sure that they were always fully alert and aware of the road environment and so only interacted with the participant at times that they felt it was safe to do so. The participant was also instructed to respond only when they felt able to.

8.2.4.4 NASA-TLX

The NASA-TLX (Hart & Staveland, 1988) was administered to participants after the road trial and the simulator trial. Participants were required to retrospectively rate their workload for when they were asked to respond to the scenarios. The subjective workload of the participants when providing verbal responses in the simulator could then be compared to the road.

Willingness to Engage with Technology

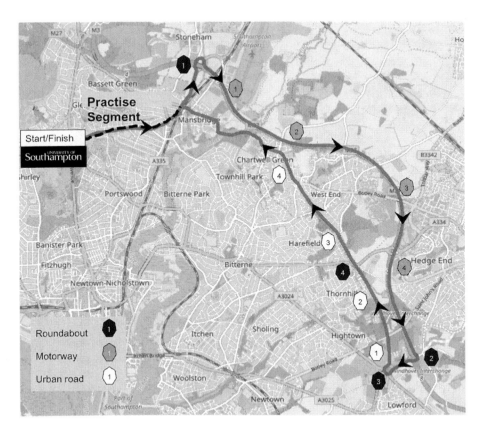

FIGURE 8.2 Map of the route showing the locations that the scenarios were posed and the practise segment.

8.2.5 Data Analysis

Qualitative data was obtained by transcribing the recordings from the participants' verbal reports, these were then input into NVivo 11 for further coding and analysis. The drivers' intention to engage was coded as 'yes' or 'no', in line with the drivers' reported decision in their verbal protocol. The verbalisations of the drivers' interpretation of the current road environment and how this influenced their intention to engage were coded with the thematic framework developed in Chapter 6 (Parnell et al., 2018a). This is included in Appendix A of this book. The use of NVivo 11 to code the qualitative data also facilitated quantitative analysis to be run on the frequency of references made to the nodes of interest. To determine if responses in the simulator were consistent with responses on the road, Matthews (1975) correlation coefficient (Phi) was calculated. Phi is typically used in machine learning as it calculates the correlation between predicted and observed binary outcomes. The measure has been used to assess the validity of a range of popular Human Factors methods (Stanton & Young, 1999). It was applied here to determine the correlation between the frequency of yes/no responses reported in the simulator and the road condition. An interaction between the response given (yes or no), the road type and task was explored using a three-way loglinear analysis. The analysis aimed to identify if drivers were more likely to respond one way or another on specific road types or in relation to different tasks as well as assessing any interacting effects between these factors.

The number of references made to the thematic framework were also correlated between the road and the simulator condition. This assessed the relationship between what drivers rate as important to their decision-making process when engaging with technology on the road and in the simulator. A Spearman's rank correlation was conducted to correlate the frequency of responses to each of the thematic themes across both conditions.

Quantitative data was also obtained from other metrics, including the drivers' workload, average speed and speed variability while driving, for statistical analysis. Paired samples t-tests were run on the drivers' rated workload in the simulator and on the road. Measures of the drivers' average speed and variation in speed when driving with and without providing verbal protocols on their intention to engage with technology were obtained from the simulator and on the road. Speed metrics could only be obtained and compared on the motorway and A road, not the roundabouts, as breaking inputs could not be generated from the IV.

The motorway and A road were split into segments where the drivers were verbalising their intention and where they were driving without verbalising their intention. Periods of comparable speed limit and road curvature were compared across the route for the analysis of speed when they were providing verbal reports and when they were not. Areas where external sources may have influenced the drivers' speed were removed, such as intersections on the A road and junction exits on the motorway, as well as 100 m before and after these points (Salmon, Walker et al., 2017). The speed on the motorway and A road were compared separately to assess if there were any differences in the metrics between these road types. Comparisons between the road and the simulator could not be conducted due to the different sampling rates with which the speed data was obtained during these trails. Speed was not found to be normally distributed in the simulator condition, therefore Wilcoxon signed rank tests were performed to assess the effect of performing verbal protocols on the A road and the motorway. In the road condition, the data was normally distributed and paired sample t-tests were used to assess average speed on the A road and motorway. Again, speed variability in the simulator was assessed with a non-parametric Wilcoxon signed rank test, whereas paired sample t-tests were run on the road data to compare the effect of verbal protocol on speed variability across A roads and motorways.

8.3 RESULTS

Assessment of the drivers' intention to engage with in-vehicle technologies in a variety of different scenarios while driving in the simulator and on the road is presented. Assessment of the drivers' metrics also allowed the effect of providing verbal protocols on the drivers' speed.

Willingness to Engage with Technology

8.3.1 SCENARIO RESPONSES

8.3.1.1 Pre-Trial Interview

The pre-trial interview assessed the different ways that the participants stated they would interact with the tasks presented to them in the scenarios. Numerous ways of responding to the technology were suggested in the pre-trial interview, these are shown in Table 8.4.

8.3.1.2 Stated Intention

Intention to engage with technologies in the simulator environment was compared to intention in the real world using Matthews (1975) correlation coefficient (Phi). The frequency of yes/no responses in the simulator and on the road were calculated and classified as follows:

Hit: Simulator = 'yes' and Road = 'yes'
Miss: Simulator = 'no' and Road = 'yes'
False Alarm (FA): Simulator = 'yes' and Road = 'no'
Correct Rejection (CR): Simulator = 'no' and Road = 'no'

Calculation of Matthews (1975) correlation coefficient returned a phi value of 0.68. The figures for this calculation are given in Table 8.5. This suggests a strong positive correlation between drivers' reported intention to engage with the technology in the simulator and on the road.

TABLE 8.4

Ways Participants Stated They Would Interact with the Tasks in the Scenarios Posed in the Pre-Trial Interview

Task Scenario	Type of Interaction
Read a text	Read who it was from
	Glance at the first line
	Open the text
	Enter pin code to read
Change a destination	Recent destination
	Enter post code manually
	Voice command
Make a phone call	Via the phone
	Via the hands-free system
	Use voice command to initiate
Change a song/radio station	Steering wheel button
	Central console button
	Phone

160　　　　　　　　　　　　　　　　　　　　　　　　Driver Distraction

TABLE 8.5

2 × 2 Contingency Table of Responses on the Road and in the Simulator

		Road	
		Yes	**No**
Simulator	Yes	(Hit) 77	(FA) 5
	No	(Miss) 18	(CR) 44

8.3.1.3　Road Type and Task Type

As the frequency of responses across the simulator and the road were found to have a strong positive correlation, the percentage of 'yes' and 'no' responses in both the simulator and the road conditions were calculated. These are shown in Table 8.6.

A three-way loglinear analysis produced a final model that retained two-way effects. The highest order interaction (task*road*responses) was not found to be significant [$X^2(6) = 1.49$, $p = 0.96$]. The level two interaction was significant [$X^2(17) = 41.22$, $p < 0.01$]. To break down this effect, separate chi-square tests were run for responses to task type and road type. For task type, there was a significant association between task type and responses (yes or no) [$X^2(3) = 13.77$, $p < 0.05$]. Odds ratios suggested that participants were 3.04 times more likely to change a song than make a phone call, 2.56 times more likely to change a song than read a text and 3.60 times more likely to change a song than changing a destination on satnav. The direction of

TABLE 8.6

Percentage of 'Yes' and 'No' Responses Made across Road and Task Type in the Simulator and Road Conditions

		% of 'Yes' Responses	
Road Type	**Scenario**	**Simulator**	**Road**
A road	Phone	50	75
	Text	58.33	75
	Destination	50	66.67
	Song	91.67	91.67
Motorway	Phone	66.67	66.67
	Text	75	66.67
	Destination	66.67	66.67
	Song	83.33	91.67
Roundabout	Phone	41.67	33.33
	Text	33.23	50
	Destination	25	33.33
	Song	41.67	75

Willingness to Engage with Technology

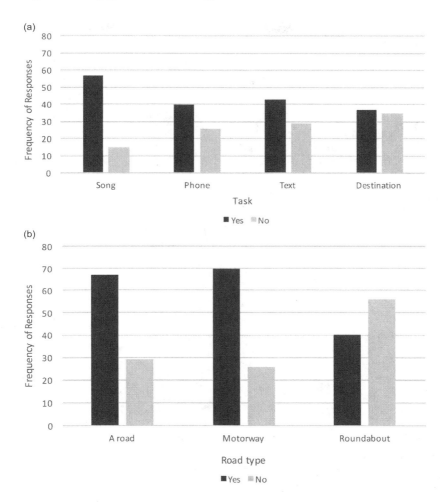

FIGURE 8.3 Frequency of responses across (a) task type and (b) road type.

these differences can be seen in Figure 8.3a. For road type, there was significant association between road type and response [$X^2(2) = 24.01, p < 0.01$]. The odds of drivers responding 'yes' to wishing to engage with a task on a roundabout was 3.77 times lower than on the motorway and 3.24 times lower than on an A road (see Figure 8.3b).

8.3.1.4 Reasons for Stated Intention

While the drivers' intentions were found to have a strong positive correlation in the simulator and on the road, it was also of interest to assess if the drivers gave the same reasoning for their intentions across both research settings. The percentage of references that were coded to each of the themes from the previously generated thematic framework are shown in Table 8.7.

Table 8.7 shows that a theme has been added to the original framework that is presented in Table 8.7, 'Other road users'. The coding procedure revealed that there were

TABLE 8.7

Percentage of References Coded to the Thematic Framework during the Verbal Reports Made in the Road and Simulator Conditions

Key Themes	Percentage of References	
	Road	Simulator
Context	**6.7**	**5.6**
Journey	1.9	1.9
Road	1.3	0.3
Task	3.5	3.4
Driver	**5.3**	**5.8**
Attitude of the driver	4.7	4.5
Influence of others	0.2	0.3
Tendency	0.2	0.0
View of self	0.2	1.0
Infrastructure	**38.9**	**38.5**
Illegality	0.0	0.3
Perception of surrounding environment	9.6	9.1
Road layout	5.5	5.7
Road-related behaviour	11.5	11.2
Task-road relationship	12.3	12.2
Task	**21.7**	**21.6**
Ability to complete	0.8	0.5
Complexity	4.6	2.8
Desirability	0.2	0.7
Duration	3.8	2.4
Engagement regulation	4.4	5.7
Interaction	7.9	9.5
Other road users	**27.8**	**28.4**
Car	11.9	10.8
Behind	3.6	1.7
In front	3.5	3.4
On coming	0.5	1.2
Overtaking	0.2	0.9
Parked	0.3	0.7
Predictability	1.4	2.1
Speed	0.8	0.5
Turning	0.8	0.7
Cyclist	0.3	0.0
Lorry	1.3	1.5
Overtaking	1.3	1.5
Pedestrian	1.7	2.9
Children	0.3	0.3

NB the additional theme 'Other road users' and its subthemes are outlined as they are new to the framework. Bold text added to signifies the different clustering.

Willingness to Engage with Technology

a significant number of meaningful references to other road users that influenced the drivers' intentions to engage, with other cars, lorries, cyclists and pedestrians frequently mentioned. Therefore, a new theme and its corresponding subthemes were added to the framework to capture these references. The subthemes were coded inductively through the same process used to generate the original framework in Chapter 6 (Table 6.2) and detailed in Parnell et al. (2018a). The descriptive themes (e.g. behind, in-front, on-coming) are shown in Table 8.7 to indicate what the subthemes are referencing. Analysis of the remainder of the results will focus on the semantic subthemes only.

It can be seen from Table 8.7 that the percentage of references to each of the key themes in the road condition was similar to the percentage of references in the simulator condition. A Spearman's rank correlation coefficient showed a very strong positive correlation between the frequency of references to the sematic themes given for engaging or not engaging with the technology in the simulator and on the road $[r_s(20) = 0.93, p < 0.01]$. This suggests that not only were the responses similar in the simulator and on the road, but the reasons that drivers gave for their answers were highly correlated as well.

8.3.2 MATRIX QUERIES

The loglinear analysis suggested that drivers were less willing to engage with the tasks on roundabouts in contrast to the motorway and A road. They were also more likely to respond to the song change scenario than the other technological scenarios. The coded transcripts from the verbal reports were analysed to explore why this may be the case. A matrix coding query was run in NVivo 11 to revel the frequency of references made to each of the themes in the framework across each of the tasks (Table 8.8) and each of the road types (Table 8.9)

Change a song: Table 8.8 shows that the song scenario had fewer references to the infrastructure themes than the other scenarios and more references to the task themes. The task references that were made in response to the song scenario reflected a unanimous agreement that it was the ease of completing the task which increased their propensity to engage with it across road types.

> I'd probably glance once and then I would just know what I'm doing. I can feel the buttons so I'd be quite – I'd have no problem doing that.

> *(Participant 1)*

> That again is because it is a quick action to do in my car.

> *(Participant 2)*

There was less focus on the road type and the infrastructure for the song task as drivers claimed it did not influence their decision.

> Even when diving across lanes, probably could change a song quite quickly.

> *(Participant 3)*

> Yeah I would change that now, straight away. I probably wouldn't spend much time assessing the situation I would probably just reach out and change it.

> *(Participant 5)*

TABLE 8.8
Frequency of References to the Themes in the Framework across Each of the Scenarios

Themes	Destination	Call	Song	Text
Context	21	21	9	24
Journey	7	7	3	6
Road	3	1	3	3
Task	11	13	3	15
Driver	11	18	13	25
Attitude of the driver	8	12	12	24
Influence of others	2	0	0	1
Tendency	1	0	0	0
View of self	0	6	1	0
Infrastructure	121	108	86	104
Illegality	1	1	0	0
Perception of surrounding environment	24	32	27	31
Road layout	12	18	18	20
Road-related behaviour	40	37	23	38
Task-road relationship	56	37	26	30
Task	42	60	69	55
Ability to complete	1	3	1	3
Complexity	12	15	13	5
Desirability	2	0	3	0
Duration	10	7	15	6
Engagement regulation	6	11	8	36
Interaction	16	36	42	11
Other road users	28	31	30	36
Car	29	32	35	38
Cyclist	0	0	1	1
Lorry	4	3	4	6
Pedestrians	5	11	5	7

Whereas for the other tasks, drivers talked more about how they may (or may not) integrate their technological interactions with the technology with the current road environment, referencing the 'task-road relationship' theme.

> I would wait until I've kind of done… well, I'd either kind of slow down in this lane or I'd wait to do my manoeuvre.

> *(Participant 2)*

> I would do the voice entry, but I wouldn't want to take my hands off the wheel right now because it's quite a windy road and the guy in front of me is quite slightly erratic, so I'd try the voice thing but I wouldn't try to on a touch screen.

> *(Participant 11)*

Willingness to Engage with Technology 165

TABLE 8.9

Frequency of References to the Themes in the Framework across Each of Road Types

Key Themes	A Road	Motorway	Roundabout
Context	28	20	27
Journey	4	4	15
Road	5	4	1
Task	19	12	11
Driver	17	25	25
Attitude of the driver	11	24	21
Influence of others	2	0	1
Tendency	0	1	0
View of self	4	0	3
Infrastructure	120	123	176
Illegality	1	1	0
Perception of surrounding environment	37	41	36
Road Layout	36	16	16
Road-related behaviour	38	33	67
Task-road relationship	21	47	81
Task	85	86	55
Ability to complete	1	4	3
Complexity	18	20	7
Desirability	3	2	0
Duration	15	10	13
Engagement regulation	21	23	17
Interaction	38	44	23
Other road users	200	80	42
Car	76	36	22
Cyclist	2	0	0
Lorry	3	13	1
Pedestrians	26	1	1

Read a text: There were proportionally more references to the 'engagement regulation' theme under the task category for the 'reading a text message' scenario. These references relate to statements suggesting they may initially be drawn to the text upon receiving it and may read who it was from before determining if they would read the whole text. They suggest how the presentation of the text would initially guide their attention to it, and they would assess how they would then regulate their engagement with it.

> Once I've had a look at who the text message was from, I would assess whether I wanted to look at it or not, depending on how important the person was to me.
>
> *(Participant 3)*

> I'd read like the first line of it, but I probably wouldn't open the text up to read the rest of it unless it was all shown on the front anyway.
>
> *(Participant 11)*

166   Driver Distraction

Change a destination: This task had the most references to the infrastructure themes, which was also the case in the interview study where it was stated to be due to the use of the satnav to navigate the road environment and therefore there was a close association between the two. Notably, in this study, there were a large proportion of references to the 'Task-road relationship' subtheme which related to drivers stating their need to change a destination would vary on their familiarity with the road environment. While others stated that they would be able to pull over on the A road and would therefore not do it while driving. On the motorway, drivers stated it would depend which lane were in, with many suggesting they would do it in the slow lane.

Yep, so certainly around here where I don't know the area very well.

(Participant 5)

I wouldn't, necessarily, enter a destination into my sat nav now, because – well, yeah, I mean I can pull over relatively easily, and the distance I'd cover in that time in waiting to find somewhere to pull over would be relatively small.

(Participant 2)

I would do it on the voice input, if that didn't work, then I'd probably wait for a, sort of, quieter bit, and pull into the slow lane.

(Participant 11)

Make a phone call: The task of making a phone call is one of initiating rather than responding, which drivers stated they would be less likely to engage with in the interview studies. Yet, interview subjects did suggest that they would make a call if they needed to, which is why this scenario was posed. It was left open for the drivers to state if they used hands-free or other methods to make the call, which some did, although not all. Despite the illegality of the task only one reference was made to this subtheme. Although the task did have the most references to the 'view of themselves' subtheme which suggested they would feel they should could not perform the task.

I wouldn't feel comfortable doing that.

(Participant 10)

Because I don't think I'm good enough at concentrating for speaking and look and indicate.

(Participant 11)

Many drivers stated they would never use their phone while driving and would wait to make the phone call when they arrived. Others stated the complexity of the phone interface would stop them interacting. While others stated they utilised hands-free devices to allow them to make calls.

No, I wouldn't. I'd wait until I'd got there, and catch up with phone calls.

(Participant 10)

Because I don't want to scroll through menus and click things and repeatedly look and look and look.

(Participant 4)

Willingness to Engage with Technology

167

> Yeah, so this is connected, but if not, I would be happy to connect my hands free by putting the Bluetooth on my phone.
>
> *(Participant 2)*

A road: When assessing references to the themes by road type, Table 8.9 shows that the A road had a larger number of references to the 'other road users' theme than the motorway or the roundabout road type. When driving on this road, participants' verbal reports were strongly influenced by the presence of others in the road environment including pedestrians as well as the unpredictable nature of the A road. This was in contrast to the motorway that was perceived as a more stable environment.

> Yeah, so there's no cars behind me, there's no pedestrians, it's quite quiet, there's a few cars coming towards me.
>
> *(Participant 1)*

> Yep so a massive gap to the car in front, there's absolutely nothing has followed me from the roundabout behind, so yeah I could flick through the radio stations to my heart's content.
>
> *(Participant 3)*

Roundabout: The roundabout road types, which had significantly fewer 'yes' responses, had more references to the infrastructure and less references to the task than the A road and motorway. The frequent references to the 'task-road' relationship suggest that the attention required when navigating a roundabout limited their ability to engage with a secondary task.

> I'm waiting at the roundabout because there's a car in front of me which has just gone, so I'm looking at whether I can go. And I'm looking at the roundabout and getting onto the roundabout at the moment. No, I've got to do too much manoeuvring of the car and thinking about the environment to bother looking at my phone.
>
> *(Participant 1)*

Motorway: On the motorway, references to the infrastructure themes frequently mentioned the road environment and the 'road-related behaviour' theme, including the behavioural differences across the motorway lanes that influenced their willingness to engage.

> Probably there I would have delayed slightly until I was in the right lane, the left lane.
>
> *(Participant 3)*

> I'd stay in this lane. I wouldn't try and change lane while I did it, and I'd try and keep... leave like a good distance between me and whatever was in front and just try and keep an eye out for someone pulling into that gap while I was on the phone.
>
> *(Participant 11)*

> Again, I'd pull into the slow lane, and then I'd probably do it now, yeah.
>
> *(Participant 12)*

8.3.3 DRIVING SPEED

Measures of the drivers' average speed and variation in speed when driving with and without providing verbal protocols on their intention to engage with technology were obtained for the simulator and the road condition. The speed on the motorway and A road were compared separately to assess if there were any differences in the metrics within these road types. The average speed and the standard deviation of speed are presented in the box plots in Figures 8.4 and 8.5, respectively. Outliers were categorised as those 1.25 standard deviations away from the mean. The descriptive data is shown in Table 8.10.

8.3.3.1 Mean Speed

No differences were found between the drivers' mean speed in the simulator when they were providing verbal responses compared to when they were driving without verbally responding on either the A road [$Z = -0.55$, $p = 0.58$), $r = -0.11$] or the motorway [$Z = -1.10$, $p = 0.272$), $r = -0.22$]. There were also no significant differences in mean speed when verbalising responses compared to driving without verbalising responses on the motorway in the on-road condition [$t(11) = -0.16$, $p = 0.876$, $r = 0.89$]. There was, however, a significant difference on the A road when drivers

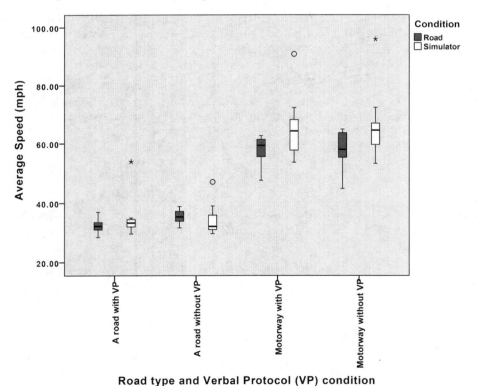

FIGURE 8.4 Box plot to show the differences in the average speed across the motorway and A road when driving with and without VP. The road condition and simulator condition are presented.

Willingness to Engage with Technology

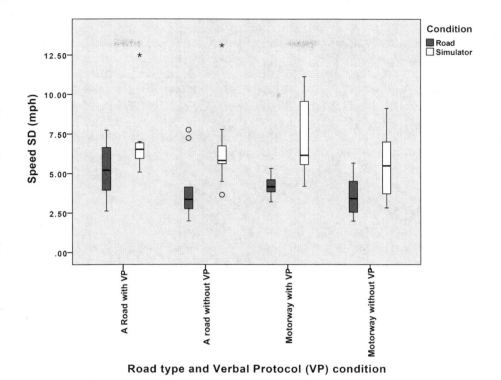

FIGURE 8.5 Box plot to show the differences in the standard deviation of speed across the motorway and A road when driving with and without VP. The road condition and simulator condition are presented.

TABLE 8.10
Mean Speed and Speed Variability for the Road and Simulator Condition

	A Road		Motorway	
	Verbal Protocol	No Verbal Protocol	Verbal Protocol	No Verbal Protocol
Road Condition				
Average speed	32.39(1.87)	35.46(1.77)	58.15(3.78)	58.29(5.15)
Speed variability	3.93(1.82)	5.33(1.65)	3.62(1.16)	4.22(0.60)
Simulator Condition				
Average speed	34.11(1.70)	34.72(2.74)	62.81(5.92)	62.89(5.55)
Speed variability	6.83(1.88)	6.42(2.36)	7.28(2.33)	5.55(1.99)

were driving on the road, with an increased average speed when driving without providing verbal reports (M = 35.46, SD = 2.74), compared to when they were providing verbal responses (M = 32.39 mph, SD = 1.87), [$t(11) = -4.77, p < 0.01, r = 0.44$].

8.3.3.2 Speed Variability

In the simulator condition, there was no significant difference in speed variability on the A road when providing verbal protocol (VP) compared to when they were not [$Z = -1.26$, $p = 0.21$, $r = -0.26$]. Yet, there was a significant difference when they were driving on the motorway in the simulator, with increased variability when providing verbal responses (M = 7.28, SD = 2.33) compared to driving without verbal responses (M = 5.55, SD = 1.99), [$Z = -2.67$, $p < 0.01$, $r = -0.54$]. In the road condition, there was no significant difference in speed variability on the motorway [$t = -1.198$, $p = 0.081$, $r = -0.38$] or the A road (after correcting for the multiple test with a Bonferroni correction) [$Z = -2.20$, $p = 0.028$, $r = -0.45$].

8.3.4 WORKLOAD

An overall workload score was obtained for the simulator and road condition by averaging the six subscales of the measure (mental, physical, temporal, performance, effort, frustration) (e.g. Horberry et al., 2006). Workload ratings were found to be significantly higher in the simulator condition (mean = 8.41, SE = 0.87) than in the road condition (mean = 5.61, SE = 0.70), [$z = -2.82$, $p < 0.01$, $r = 0.58$]. This is shown in Figure 8.6.

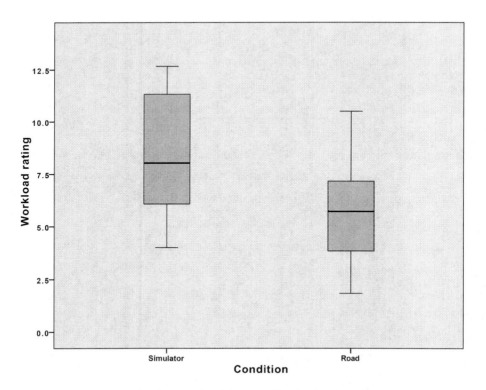

FIGURE 8.6 Boxplot showing the workload scores for the simulator and road condition.

8.4 DISCUSSION

Noting that distractions are largely voluntarily engaged by the driver has prompted the need to assess the decisions that drivers make and the factors that influence their engagement in distracting tasks. The assessment of decision making in its natural environment was highlighted as a useful, but under-researched, method to assess this (e.g. Horrey & Lesch, 2009; Schömig & Metz, 2013). The drivers' decision to engage was assessed in a simulator setting and on the road to determine what influences the decision and the reasoning given by drivers.

8.4.1 EXPERIMENTAL CONDITION

A strong positive correlation was found between the drivers' decision to engage in secondary tasks, both in the simulator and on the road. Furthermore, the key themes that drivers referenced in their reasoning for their intention were also strongly correlated on the road and in the simulator. This provides relative validation of the simulation developed for the study, as well as showing promise for the use of simulation to observe, measure and influence the drivers' intention to engage with technological devices of distraction for future research.

Furthermore, the application of the inductively generated thematic analysis from the interview study in Chapter 6 (Parnell et al., 2018a) to the drivers' verbal reports when driving in a simulated and real-world environment suggests that participants are able to generate realistic decision-making processes in an interview setting. The addition of the 'other road users' theme found that drivers were able to comment more on their interactions with other actors in the road environment (both simulated and on-road) which could not be anticipated in the interview setting. This highlights the benefits of assessing the validity of the framework in the real-world environment to increase its ecological validity. It also compliments other decision-making studies that have found driver behaviour to be influenced by other drivers (Pedic & Ezrakhovich, 1999). Yet, it diverges from the findings of the naturalistic study conducted by Tivesten and Dozza (2015) who observed that drivers' decisions to engage with visual-manual tasks was not influenced by the presence of lead vehicles. The drivers' verbal responses in this study suggests that drivers are actually very aware of the other vehicles on the road surroundings.

8.4.2 FACTORS AFFECTING NATURALISTIC DECISION MAKING

Analysis of drivers' stated intention to engage across road type and task type showed evidence for strategically adapted engagement. This supports the findings of Schömig and Metz (2013) in their simulator study and Tivesten and Dozza (2015) in their naturalistic driving study, but opposes the lack of strategical adaption found by Horrey and Lesch (2009) on the test-track.

The frequency of themes that were coded to the drivers' verbal responses in Tables 8.8 and 8.9 highlighted how the themes may vary in their influence over the driver decision to engage across tasks and road types, respectively. Acknowledging that the drivers' decision to engage with possible distracting technologies is dependent

172   Driver Distraction

on a large range of factors that are likely to vary across different scenarios is very important. It is key that road and vehicle designers understand how their choices may impact on the drivers' decision to engage with tasks that may lead them to become distracted. In Appendix B, Tables B.1 and B.2 are intended to summarise to the key factors that influence the decision to engage across tasks and road types. These highlight both the factors that are likely to increase the drivers decision to engage and those that would decrease their likelihood of engaging. As the reports were not found to differ across the road and simulator conditions, these factors were drawn across experimental environments and therefore can also be applied to both environments. These tables may be of interest to road and vehicle designers in order to guide their own decisions on what facilities they provide the driver and how they can shape their behaviour. Further discussion around these points are presented below for both task type and road type.

8.4.2.1 Task Type

In line with the findings in the interview study, drivers were two to three times more likely to respond to the 'changing song' scenario than the scenarios requiring 'phone', 'text' or 'destination change' responses. This supports previous findings from survey studies asking drivers to rank tasks on their potential for distraction, where in-car entertainment systems and radio controls are typically ranked lower than phone-based or satnav interactions that are perceived to be more distracting (Young & Lenné, 2010; Lansdown, 2012). The verbal reports that were coded to the thematic framework suggest that drivers were less concerned with road type when deciding whether to engage with the song task but were motivated by the ease of the task and the limited attention it took away from the driving task. This mirrors the discussions in the interview study and supports previous findings by Lerner (2005) in the real-world environment. Intention was lower for other tasks (phone, text and navigation), with issues frequently referencing the complexity of the task. Yet, where drivers did state they would engage, they reported their behaviour would involve several steps that would lead them to approach the task in a staged manner while adapting their driving behaviour. For example *"if I had my sat-nav kind of up here I could start to slowly enter the destination and just kind of maybe doing one step at a time and then looking back to the road to make sure nothing's changed"* (Participant 3). These findings suggest that, once drivers have made the strategic decision to engage with the task, they adapt their behaviour on a momentary basis to integrate the primary and secondary tasks, a feature of adaption at the decision and control level (Schömig & Metz, 2013). Harvey and Stanton (2013) found that drivers were able to use 'shared glances' to monitor both the road and a visual display at the same time once a task had been engaged with, in order to perform the two tasks concurrently. The use of shared glances was very apparent when entering destinations into a navigation display (Harvey & Stanton, 2013). The behaviour drivers discussed in their verbal reports, such as the comment made by Participant 3 above, was suggestive of this 'shared glance' model. Thus, drivers are aware of their behavioural adaption at these lower levels of adaption as declared by Schömig and Metz (2013).

The scenario that asked drivers if they would read a text that they had just received, frequently referenced the 'engagement regulation' theme which captures

Willingness to Engage with Technology

references to how the initiation of a task is regulated. This is due to the number of participants who stated that they would read a text message if they were alerted to it first, to see who it was from and then assess if they wanted to go on to read the text in full. Others claimed their device would allow them to read the first few lines without touching the phone, which would then allow them to assess if they would go into the phone to read the full message. This reflects the resolution of conflicting tasks through assigning a small amount of resources to determine the secondary tasks priority. This reiterates the connection between 'resource constraints' and 'goal priority' that was not evident in the original PARRC model but was found when the thematic framework and the interview data was applied to the PARRC model in Chapter 6 (Parnell et al., 2018a). While reading the first few lines of the text would allow them to decide the priority of the text, it would also increase the length of time that drivers would need to take their eyes away from the road in contrast to reading just the name (Harvey & Stanton, 2013). This furthers previous statements that mobile phone manufacturers may be influencing the drivers' diversion of attention away from the road, re-surfacing the issue of prohibiting the driver from interacting with the mobile phone, but permitting the mobile phone to interact with the driver (Parnell et al., 2018b).

8.4.2.2 Road Type

The finding that drivers were approximately three times more willing to engage with a task on the motorway or A road than at roundabouts supports previous findings that suggest drivers are less willing to engage in other tasks while performing manoeuvres (Lerner et al., 2008). It also reiterates the findings from the previous chapter that report increased likelihood of engaging when stopped at traffic lights in comparison to when driving through a junction (Chapter 7). Whilst drivers seemed indifferent in their intention to engage on A roads or motorways, the relationship between the task and the infrastructure at roundabouts was mentioned frequently when drivers were at roundabouts. The large amount of processing that was required to navigate the roundabout was mentioned by many participants (e.g. *"At the moment, no, because we're on a junction. There's a lot going on. There's things in all the different lanes. And if it was any more than just one press, that would be too much for me to do"*, Participant 4). This suggests that drivers are able to assess when they are unable to engage with technological tasks while driving and avoid doing so at complex junctions, providing further evidence for their strategic adaption of task engagement.

While drivers were more likely to engage on motorways and A roads, they did vocalise features of these environments that would influence their decision to engage. On the A road, drivers were notably more concerned with the presence of other road users including pedestrians who may step out in the road and other cars who may act unpredictably. On the motorway, drivers made multiple references to the other cars around them, the lane of traffic that they were in and how this would influence their decision to engage, with many stating that they would pull into the slow lane if they needed to engage in a non-driving task concurrently with the driving task. There was also a number of references to lorries and large vehicles that they needed to overtake on the motorway that influenced their decision. This suggests that the drivers did incorporate the surrounding environment into their strategic planning of secondary

174 Driver Distraction

task engagement. Furthermore, not only was strategic intention evident across road types but also within them. The role of road infrastructure design is thus revealed in the drivers' decision to engage with distractions.

8.4.3 Using Verbal Protocol to Capture Naturalistic Decision Making

The use of verbal protocols aimed to determine drivers' intention to engage, rather than assessing what happens to driving performance when the driver becomes distracted. Understanding the decisions that drivers made and the reasoning for their decisions can inform effective mitigation strategies to target the underlying cause of distraction, rather than penalising drivers once it has happened. Verbal protocol is a simple and low-cost way of collecting data about, and analysing, cognitive processes (Ericsson & Simon, 1993). This had obvious advantages to the methodology that could be used and the settings within which it could be measured. The assessment of intention to engage, rather than making the driver physically engage, mitigated some of the safety concerns of studying the behaviour in the real world. The pitfalls and complications in the study of real-world driving behaviour are noted by Carsten et al., (2013) who highlight the difficulties in capturing natural behaviour without noise and the influence of other confounding factors.

The use of verbal protocols in the real-world driving setting have, however, suggested improvements to driving performance and safety (Salmon, Goode et al., 2017). It has been found that the continuous verbal reports that drivers provided did not influence driving speed or control, and at junctions, it actually improved the drivers' braking and acceleration (Salmon, Goode et al., 2017). These findings, however, relate to the continuous verbal reports of drivers to assess the drivers' situational awareness, rather than assessing their decision-making process. Therefore, to assess the impact of providing verbal responses to decisions, speed metrics were compared during periods when drivers were providing verbal reports, stating their intention to engage in response to the scenarios, and when they were driving without providing verbal intentions. It should be noted, however, that while speed metrics such as average speed and speed variability can infer distraction (e.g. Horrey & Wickens, 2004; Burns, et al., 2002), they are not perfect predictors and may be influenced by other aspects of the road environment and other road users. Yet, where alternative metrics such as headway cannot be obtained, as was the case in this study, they may offer some insights into driver performance (Kircher & Ahlstrom, 2010). Caution is advised in the interpretation of these findings due to the small sample size.

Nonetheless, subjective workload was found to be significantly higher when drivers were verbalising their intention to engage with non-driving tasks in the simulator compared to when doing so on the road. This is in contrast to others who have reported similar workloads on the road and in the simulator (Patten et al., 2006; Cantin et al., 2009). The debrief interviews conducted with drivers after the experimental trials revealed that many found the simulator condition difficult due to a lack of awareness of speed. Particular reference was given to roundabouts, which can be difficult to simulate and can heighten motion sickness (Bittner et al., 2002). These factors within the simulator condition may have influenced these finding. Furthermore, participants were only given a practise drive of approximately 10 min

Willingness to Engage with Technology

in the simulator before the experiment. Extensive training in the simulator environment may have reduced workload. Yet, despite increased workload, participants were able to maintain the same mean speed when providing verbal reports as when they were driving without verbalising their intention on the A road and motorway. Speed variability, however, was found to differ when driving on the motorway in the simulator, with increased variability when stating their intention to engage. This may suggest that drivers' verbal reports may have influenced their performance in the driving task at this time. One explanation for this may be due to the reduced perception of speed attributed to the motorway environment and simulator setting mentioned in the debrief interview. It may also be due to a reduced risk that accompanies simulation research (Bella, 2008) that led drivers to limit their investment in managing their speed consistently while also verbalising their intention. Indeed, some participants commented that they felt it hard to infer what their driving speed was in the simulator condition. Conversely, in the real-world condition, where the risk of accident is greater, a difference between the mean speed when driving with verbal reports and without verbal reports on the A-road was found. A roads led to frequent references to other road users, including other vehicles, pedestrians and cyclists. The complexity and increased risk of accident that this may have posed to participants in the road condition may have led the drivers to adapt their speed when verbalising their intention. Adaption at the control level has been found in other complex driving situations with drivers adapting their speed when engaging with secondary tasks to compensate for their reduced performance (e.g. Alm & Nilsson, 1994; Strayer et al., 2003).

8.4.4 FUTURE RESEARCH

Further research is required to determine how verbal protocols relate to driving speed and vehicle control on the road and in the simulator. This requires the use of a larger sample of drivers and the comparison of driving while stating intention to be compared along the same road segments as a control drive, when no verbal protocol is performed. The addition of steering and braking metrics would assist in the interpretation of this data but was unable to be collected within this study. Furthermore, the use of eye-tracking metrics to determine where the driver was looking and what they were monitoring in the moments before they declared their intention to engage would be a useful addition to this data set. Assessment of this in the simulator and the road setting would build on the findings within this study to determine how the research setting may influence the drivers' intention to engage and their decision-making process. It would also be of interest to assess how drivers may perform if they were asked to actually complete the tasks. This would assess whether the drivers' perceptions about their ability to complete the task were accurate.

A limitation of this study is the small sample size ($n = 12$). Quantitative studies conducted in simulators have afforded larger sample sizes (e.g. Jahn et al., 2005, $n = 49$; Horberry et al., 2006, $n = 31$). Yet, this research has taken an initial look at the potential comparisons that can be drawn from the simulated environment that is commonly relied upon in driver distraction research and real-world driving conditions. While the sample were of a range of ages and comprised of regular and

176   Driver Distraction

experienced drivers, the generalisations that can be made from this limited sample are restricted. Further research to needed to assess the validity of these findings and their generalisations to larger samples. Determining if the findings presented in Tables B.1 and B.2 (Appendix B) apply to a broader sample is of particular interest in order to assess the potential impact it could have on manufacturers and designers.

8.5 CONCLUSION

The decision to engage with secondary tasks has been under-researched in previous years, with the active role that the driver plays in managing their interaction with secondary tasks often ignored. This is despite evidence to suggest that distractions are voluntarily engaged with (Beanland et al., 2013). As technologies develop, in line with consumer demand, the factors that influence the decision to engage with tasks that may pose as distractions is important to consider. This chapter has shown that the drivers' decision to engage with technological tasks in the simulator is representative of their decision on the road. This shows the potential for further research into NDM in driver distraction research that can utilise the safe and controlled environment of simulators. Although it should be noted that the simulation used in this study was created to have high ecological validity, with the inclusion of salient localised objects.

The thematic framework developed within the interview study in Chapter 6 was validated and expanded on under the more realistic driving context. Furthermore, the use of verbal protocol methodology has been used to assess the complex interacting factors that influence the drivers' decision-making process when planning to engage with technological tasks. The table of factors stated in Tables B.1 and B.2 (Appendix B) provide reasons that both increase and decrease the drivers likelihood of engaging. These factors should be considered by manufacturers to acknowledge the implications of the devices that they develop to be used in the vehicle. Furthermore, this provides evidence that drivers do strategically plan their engagement with technological tasks to some extent, and that this is influenced by the task type and road type.

The next chapter will show how the findings from this study are applied to the PARRC model of distraction. This will be discussed in accordance with a review of the PARRC model and its evolution throughout the book to present how it has been developed, applied and validated through the research that has been conducted.

9 Evolution of the PARRC Model of Driver Distraction

Development, Application and Validation

9.1 INTRODUCTION

This book has been underpinned by the development, application and validation of the Prioritise, Adapt, Resource, Regulate, Conflict (PARRC) model of driver distraction from in-vehicle technology (Parnell et al., 2016). The introduction of the PARRC model in Chapter 4 was the first attempt, within the literature, to develop a model of distraction that showed the influence of actors outside of the driver. The realisation of sociotechnical factors in driver distraction is a major contribution of this book. The initial PARRC model was founded within the literature using a grounded theory methodology (Chapter 4). Across the subsequent chapters of the book, different methodologies have been applied to the study of driver distraction to capture the behaviour in the context of the wider sociotechnical system. This has led to the modification of the original PARRC model and further insights that the model has generated.

This chapter will provide a reflective discussion on the evolution of the PARRC model of distraction to show how the methodologies applied within the book have shaped the model and facilitated a novel understanding of the phenomenon of driver distraction. It will also suggest how the model is validated through triangulation, which requires the application of theory to more than one data source, method or investigator to assess their convergence (Hignett, 2005). This approach included four research methods that have featured in the evolution of the model: grounded theory, AcciMap analysis, semi-structured interviews and a driving/simulator study.

Chapter 3 discussed the influence that the methods employed to study driver distraction have on the data that is obtained and therefore the insights into the behaviour that can be gleaned. The review of methods that have been, and can be, applied to driver distraction in Chapter 3 suggested methods are required that enable the wider system within which the behaviour occurs to be observed, including the role of context, other actors and the interaction between them (Parnell et al., 2017b). Rather than reviewing the behaviour in isolation, the methods employed in this book have strived to adhere to the guidance given in Chapters 2 and 3, and explore driver distraction in relation to the complex system in which it occurs.

177

The main purpose of developing a model and reviewing driver distraction from a sociotechnical systems approach is to gain an insight into the effectiveness of the current methods of mitigating the behaviour. This chapter will reflect on the application of research methods used in this book alongside their development, application and validation of the PARRC model to show how they provide recommendations for future mitigation practises. The evolutionary stages in the PARRC model development, the methods that were applied and the findings that they have presented are shown in Table 9.1.

TABLE 9.1
Stages in the Evolution of the PARRC Model

Stage	Chapter	Research Method	Evolutionary Stage	Findings
1	4	Grounded theory	Development	The grounded theory methodology explored the literature on driver distraction and established five key factors of distraction and their interconnected relationships.
2	5	AcciMap analysis	Application	An AcciMap analysis of the legislation on distraction from handheld mobile phones compared to other technologies in the vehicle, applied the PARRC factors to the systems actors and actions in the AcciMaps. This utilised the factors in a similar way that other systems methods use failure taxonomies (e.g. STAMP, Leveson 2004 or HFACS Shappell & Wiegmann, 2003). This showed which actors were supporting the factors of distraction, and those which were not.
3	6 and 7	Semi-structured interviews	Validation and development	The interviews allowed the drivers to discuss their reasons for engaging with technology while driving. An inductive thematic analysis presented a thematic framework of driver-informed reasons for engaging with distractions. The application of this framework to the PARRC model validated the framework, whilst also expanding knowledge on the factors and connections between them.

(Continued)

Evolution of the PARRC Model

TABLE 9.1 (*Continued*)
Stages in the Evolution of the PARRC Model

Stage	Chapter	Research Method	Evolutionary Stage	Findings
4	8	Driving study (simulator and road conditions)	Validation	An experiment comparing the drivers' decision to engage with technological tasks on the road and in a simulator collected verbal protocol reports. The thematic framework developed in the interview study was used to code the verbal reports. Evidence for the themes was found in both experimental settings, and a strong positive correlation was found between them. The framework was expanded to include the 'other road user' systemic theme. The application of themes to the PARRC factors identified the prevalence of the factors and the connections between them. This suggested similar connections as those made in the interviews with additional context-based insights.

In addition to reviewing the evolution of the PARRC model, this chapter will also suggest how the work conducted in this book and the insights that have been obtained in the development of the PARRC model can be used to inform a novel definition of driver distraction from in-vehicle technology that is able to account for the wider sociotechnical system within which it occurs. It was highlighted in Chapter 2 that there was no universal definition of driver distraction, despite the many attempts that have been made to develop a definition of the phenomenon (e.g. Lee et al., 2008; Regan et al., 2011). Commonalities across definitions were identified, including the 'diversion of attention' implicit within distraction as a general concept and the adverse effect that this diversion of attention may have on the drivers' performance in the main goal of arriving at the destination safely. Yet, the definitions also varied on their views on the source of distraction, its intentionality and location (Lee et al., 2008). While the focus of many definitions was on the impact of the driver and their responsibility for causing distraction (e.g. Patten et al., 2004; Pettit et al., 2005), some also suggested that it may be the object or event that may be the source of the distraction (e.g. Drews & Strayer, 2008; Manser et al., cited in Young et al., 2008). Yet, no definition was able to account for the role and responsibility of the wider sociotechnical system in the emergence of distraction-related events. In this chapter, a definition of distraction is presented that is informed by the work conducted in this book that aims to incorporate the sociotechnical systems approach.

9.2 DEVELOPING THE PARRC MODEL

To determine how the PARRC model evolved through the research conducted in this book to inform the definition, each of the four stages outlined in Table 9.1 are briefly discussed in turn. The different iterations of the PARRC model as it was developed across the book are presented in Figure 9.1.

9.2.1 STAGE 1. GROUNDED THEORY: MODEL DEVELOPMENT

The lack of a universal definition and model with which to review driver distraction was highlighted in Chapter 2. While multiple theories of driver distraction were identified, their views on the occurrence and processes involved in distraction differed and did not allow for an insight into the sociotechnical systems surrounding the behaviour. The absence of a model that accounted for the role of the wider system was noted and the importance of filling this gap led to the origins of the PARRC model of distraction. This initial version of the model is shown in Figure 9.1a.

Grounded theory methodology was used to develop the model due to its ability to explore the underlying principals and causal factors that resulted in the phenomenon through studying the literature that it originates from (Glaser & Strauss, 1967; Rafferty et al., 2010). The application of the PARRC model to the case study in Chapter 4 showed that the study of the factors in isolation will not provide the full account of the behaviour or any resulting incident. Analysis of the case of Victoria McClure showed the impact of each of the factors and the interconnections between them in the emergence of the fatal incident. Furthermore, the identification of the factors in the case study allowed, for the first time, the systemic actors that influenced the emergence of the incident. While developing a model from the literature

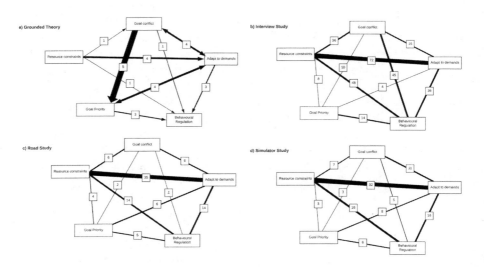

FIGURE 9.1 The versions of the PARRC model developed through data collected from (a) grounded theory, (b) an interview study, (c) a road study and (d) a simulator study.

Evolution of the PARRC Model

enables the model to have high ecological validity, grounded theory methodology can also be limiting as absent aspects of the model may not reflect absent aspects of the behaviour, they may just not have been explored within the literature. The field of driver distraction is sizable, yet filled with some contradictory theories (as discussed in Chapter 4), as well as many unknowns (e.g. Young & Regan., 2007; Regan et al., 2009). Therefore, relying on the literature as the only source of the behaviour is not advised (Hignett, 2005). Further application of the model to the behaviour was required.

9.2.2 STAGE 2. ACCIMAP ANALYSIS: MODEL APPLICATION

Stage 2 of the PARRC model evolution was its application to an AcciMap analysis in Chapter 5. The causal factors of distraction that have been realised through the development of the model, suggested their use in assessing how systemic actors may be facilitating, or preventing, distraction from occurring. An AcciMap analysis of the legislation on handheld phone use and other technology use in the vehicle was conducted. This used the PARRC factors as an alternative to the taxonomy of failures that are used in other systems methods such as the Human Factors Analysis and Classification System (HFACS; Shappell & Wiegmann, 2003) and Systems Theoretic Accident Modelling and Processes model (STAMP; Leveson 2004). The AcciMap analysis in Chapter 5 provided further support for the PARRC mechanism of distraction and how they may be implicated by the systemic actors in the socio-technical system, yet it failed to assess the interconnections of the PARRC model. Further work was required to explore the factors of distraction and their connections outside of the reports in the literature. Data collected from the drivers was therefore sought.

9.2.3 STAGE 3. SEMI-STRUCTURED INTERVIEWS: MODEL VALIDATION

It was suggested in Chapter 2 that much of the research into driver distraction has focused on the cause-effect relationship between the device, the driver and distraction-related outcomes. Yet, an understanding for how, why and when distractions occur is less well understood. Furthermore, little attempt has been made in the literature to understand this from the drivers' perspective, therefore semi-structured interviews were conducted with drivers. Drivers were encouraged in the interview setting to discuss their likelihood of engaging with distracting technological devices while driving in relation to different tasks and road types. This was conducted outside of the vehicle environment.

Whilst the PARRC model would have offered a potential theory from which to interpret the drivers' interview responses in a deductive analysis, an inductive method was chosen. The inductive analysis provided a rich insight in the data set as a whole and identified naturally occurring themes, informed by drivers themselves. As the literature from which the PARRC model was grounded in did not make any attempt to understand the drivers' perspective on the emergence of distraction, the data from the interviews aimed to capture the drivers' viewpoint. This could then be contrasted to the literature-derived PARRC model.

The semi-structured interview study provided a rich data set to explore the themes that were initially generated theoretically within the literature and contrast them to concepts that drivers deem important to their decision to engage with technology while driving. It provided validation for the causal factors of distraction in the PARRC model and extended these with the development of an in-depth thematic framework (Parnell et al., 2018a). This provided an insight into further interactions that were not included in the original model. The model is shown in Figure 9.1b. It also highlights the importance of validating theoretical models that are developed from the literature, using the self-reports of drivers.

The semi-structured interview methodology studied the interpretation of the behaviour from the drivers perspective through posing contexts that could require them to engage with devices. It could therefore understand to the interdependence of systems factors on the drivers' views, perceptions and actions. Further research was conducted to build on these findings to study the drivers' perceptions within the context that the interaction with in-vehicle technology occurs (i.e. when driving).

9.2.4 STAGE 4. DRIVING STUDY: MODEL VALIDATION

The area of naturalistic decision making was born from the realisation that decisions made in laboratories, under highly controlled conditions, are not necessarily representative of the decisions made in the real world (Klein, 1989). They are instead influenced by intuition (Phillips et al., 2004), experience (Klein et al., 1986), biases and heuristics (Kahneman et al., 1982). While the impact of devices on the drivers' attention is well documented, the drivers' intention to engage with devices while driving has been less explored, despite this forming a large aspect of the behaviour.

Examination of naturalistic decision making requires looking at the behaviour in the real world, which in this case would refer to the road environment. Yet, the developments in simulator facilities, in line with technology developments, have led to improvements in the fidelity and validity of experimental studies (Young & Lenné, 2017). The use of full vehicle simulators with realistic in-vehicle feedback and high-quality graphics can increase the validity of the simulation (Kaptein et al., 1996; Burnett et al., 2007). The influence that experimental context may have on the drivers' intention to engage with distractions required assessment. This was the crux of Chapter 8, where the drivers' stated decision to engage with technological tasks was assessed in a driving simulator and on the road. It also sought to identify the reasons that drivers gave for their decisions in both these environments and to address the relevance of this to the thematic framework developed in Chapter 6 and its application to the PARRC model of distraction.

Verbal protocol was used as a method to assess the drivers' stated intention to engage with non-driving tasks while they were driving, whilst also capturing the drivers' decision-making process and their interpretations of the surrounding environment and how this may influence their decision. This was undertaken to assess the relevance of the thematic framework to the driving context, as well as assess the influence that the experimental setting (simulator versus road) had on their assessment of the situation.

Evolution of the PARRC Model

The transcripts from the verbal protocols were coded using a deductive method informed by the the thematic framework developed in Chapter 6 to assess its validity (Hignett, 2005). Chapter 6 presented the inter-rater and intra-rater reliability tests that showed the framework could be reliably applied by different analysts as well as repeatedly by the same researcher over time. Triangulation of the framework to the application of another data source was employed with the verbal protocol transcripts.

Stated intention about secondary task engagement on the road was found to have a strong, positive correlation to their intention to engage in the simulator, as was the frequency of references made to the themes coded to the thematic framework. This provides further evidence for the utility of using a full-car, driving simulator that replicates a real-world environment to assess the drivers' intention to engage with technologies while driving. Furthermore, the thematic framework was successfully used to code the transcripts (Parnell et al., in press). Yet, there were a number of references to the influence of other road users on the drivers' decision to engage that were not evident in the semi-structured interviews and were therefore missing from the thematic framework when it was applied to the verbal protocol reports. The systemic key theme 'other road users' was therefore added to the thematic framework. The coding of this theme and its relevant subthemes that are shown in Chapter 8 (Table 8.7) were generated in a top-down fashion. The systemic theme was observed first, with the frequent references to other road users signifying the necessity of this theme. The semantic themes were then evident when reviewing the excerpts coded to the systemic theme, before the descriptive themes were realised.

The nature of the statements that were coded to this theme were associated with the 'adapt to demands' factor of the PARRC model due to the demands that were involved in managing the interactions with other road users. For example, many drivers stated they would avoid engaging until they were away from other road users:

I wouldn't read a text near a cyclist, but they've gone, there's nothing behind me.

(Participant 1)

Especially, kind of, when I'm following someone, because they could break suddenly whilst my attention's taken away from the driving.

(Participant 2)

The 'other road users' subthemes were therefore related to the PARRC model in much the same way as in Chapter 6 (Section 6.3.2). This is shown in Figure 9.2.

The coding of the verbal reports to the thematic framework, and utilising the application of the framework to the PARRC model in Chapter 6, enabled the transcripts to form a representation of the PARRC model from the on-road and simulator environments. Matrix queries were run in NVivo11 to assess the number of references that encapsulated connecting themes using the same methodology in Chapter 6. The representations of the PARRC models from the road and simulator study are shown in Figure 9.1c and d, respectively.

A Spearman's rank correlation coefficient was conducted to assess the correlation between the frequency of connections between the PARRC factors. This established a strong positive correlation between two studies ($r_s(8)=0.96, p<0.01$). This is shown

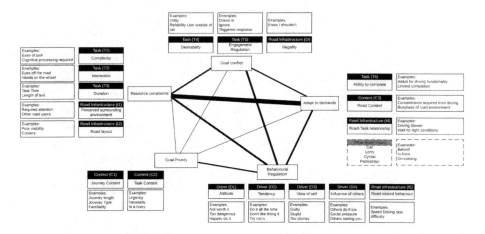

FIGURE 9.2 Diagram showing the inclusion of the 'other road users' subthemes to the adapt-to-demands mechanism on the PARRC model (outlined with a dashed line).

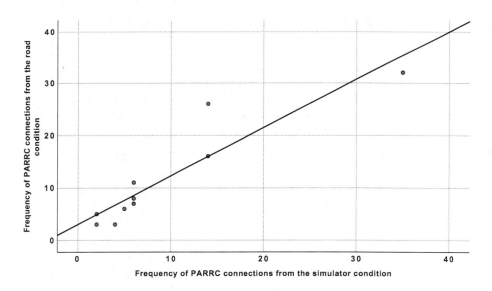

FIGURE 9.3 Scatter plot showing the relationship between the frequencies on the connections between the PARRC factors in the road condition and the simulator condition.

in the scatter plot in Figure 9.3. This further supports the relative validity of the results presented in Chapter 8.

Like the PARRC model developed from the interview study (Figure 9.1b), the findings from the road (Figure 9.1c) and simulator (Figure 9.1d) models also have a connection between 'resource constraints' and 'goal priority'. They also have strong connections from 'resource constraints' to 'adapt to demands' and 'resource constraints' to 'behavioural regulation'. There is a comparatively reduced connection

Evolution of the PARRC Model

between 'goal priority' to 'adapt to demands' and from 'goal conflict' to 'goal priority'. As suggested in Chapter 6, this may be taken to mean that drivers strive more to integrate the technological tasks with the secondary task by adapting their behaviour and regulating it in line with the resources available. This is instead of prioritising one task over another, as was depicted in the development of the original PARRC model. The findings from this stage in the PARRC model development highlight the importance of obtaining data to validate and expand theoretically developed models across different research settings.

9.3 A SOCIOTECHNICAL SYSTEMS DEFINITION OF DRIVER DISTRACTION

The development, application and validation of the PARRC model and the associated work conducted within this book has generated insights that can be used to inform a novel, sociotechnical systems focused, definition of driver distraction. Informed by the PARRC model, this definition of driver distraction states:

> Driver distraction is the diversion of attention from the main goal of arriving at the destination safely, caused by broader systemic factors at all levels (such as government policy, HMI manufacturers, media, road infrastructure, weather, traffic etc.) acting upon driver priorities, demands, resources, conflicts and constraints (identified in the PARRC model).

This systems model can be used to predict if a technology is more or less likely to be engaged with for a particular task in a given driving context. It states that distraction is not solely the result of the driver and their actions but that it is influenced by actors across all levels of the sociotechnical system. It presents the factors that influence the process through which distraction occurs; the causal factors of the PARRC model. Other definitions and theories of driver distraction have been critiqued for their inability to identify the mechanisms through which driver distraction occurs (Young & Salmon, 2012). The work conducted in this book suggests that the cause of distraction cannot be attributed to one actor. When applying the sociotechnical systems approach, a host of other actors across the hierarchical levels of the system are revealed to interact with one another to determine how safety, or conversely the potential for incident, can emerge. The PARRC factors determine how the actions and decisions that are made by systemic actors influence the potential for distraction to occur, as well as how it can be avoided. The addition of this definition to the literature is supported by the development, application and validation of the PARRC model throughout this book.

9.4 DISCUSSION

The PARRC model was born out of the findings and themes discussed in the current literature surrounding the drivers' interaction with technological devices that form sources of distraction. While this revealed much about the causal factors involved in distraction and the interacting nature of systemic actors in the emergence of system performance, primarily safety, further validation of the model was required.

Validation via triangulation of data sources allowed the PARRC model factors and interconnections to be observed within interview, simulator and road conditions. The application of findings from these experimental settings were used to expand the theoretically derived model to incorporate the drivers' perspective on the reasons they give for why they engage with technological devices while driving. Furthermore, opening the model up to the influence of systemic actors on the development of distraction has facilitated an understanding of what the systemic perspective of driver distraction should encompass and the recommendations to practise that it should propose. It has also facilitated the development of a sociotechnical systems-based definition of driver distraction, which has not yet been present within the literature. Placement of this definition alongside the other definitions of distraction that have been developed within the literature shows how it is able to account for the systemic locus of control over the emergence of distraction, as well as suggesting the causal factors that influence the development of distraction.

It was suggested in Chapter 2 that the absence of a universal definition of driver distraction has led to multiple attempts, which can make comparisons difficult and the need to apply caution when interpreting estimates that may be suggested (Lee et al., 2008; Regan et al., 2011). Therefore, the addition of another definition of driver distraction will, unfortunately, only add to this issue. Yet, a review of the current literature suggested that the need to highlight the role of the wider sociotechnical system in the initiation, development and mitigation of distraction. The addition of a systems-based definition is therefore an important addition to the research field.

The importance of understanding the reasons why distraction occurs, rather than just what happens when it does, has been highlighted throughout this book. The relevance of systemic factors to understanding the reasons for driver distraction aims to develop effective countermeasures that target the multiple causes of the issue, rather than penalising the distraction once it has already occurred. The recommendations that have been obtained through the development, application and validation of the PARRC model are now discussed.

9.5 RECOMMENDATIONS

The chapters comprising this book have provided insights into how the findings from the studies may translate into countermeasures, with recommendations and implications of the research. The emphasis within this book is on the development of systems-based countermeasures, rather than those that target the individual. The methods that have been used, and the findings presented, across the chapters offer alternative insights into such countermeasures. The notion of a research-practise gap in the mitigation of driver distraction was identified in Chapters 2 and 3 (Parnell et al., 2017b). This section will suggest how the research methods that have been applied in this book can inform strategies to implement mitigation practises in the real world.

Chapter 5 identified the potential for novel countermeasures across the hierarchical levels of the extended risk management framework (RMF) (Rasmussen, 1997; Parnell et al., 2017a). This showed how countermeasures could be applied across

Evolution of the PARRC Model

the sociotechnical system, rather than just focusing them on the end user. To reflect on the recommendations that have been suggested though the research conducted across the chapters of this book, the expanded hierarchical levels of the RMF from Chapter 5 will again be used to illustrate this point.

9.5.1 International Committees, National Committees and Government

Chapter 5 utilised the RMF in the development of the AcciMap analysis of the UK legislation and identified the role of international and national committees on the development of policy and standards at levels above the Government, at the top of Rasmussen's (1997) framework. These levels have been grouped together here, as they are in Chapter 5, as the recommendations that are suggested target all of the levels, feeding down from international committees to national governments.

In-vehicle human-machine interface (HMI) is subject to design standards and testing procedures as outlined in the NHTSA (2013) guidelines. These guidelines aim to set a universal standard for the implementation of interfaces into vehicles by manufacturers who volunteer to adopt them. They build on the previous national standards imposed by JAMA (2004), as well as the Commission of the European Communities (2008) and the Alliance of Automobile Manufacturers (2006) guidelines. These guidelines provide an outline of the procedures that should be used to test interfaces and focus on objective measurements in driving simulators which led to quantified limits on the distractive potential of the devices (NHTSA, 2013). While an attempt to provide a universal approach to these guidelines is positive, the lack of consideration for how they may be integrated with the wider sociotechnical system is evident. There is a focus on the objective/quantitative aspects of the behaviour, which were highlighted in Chapter 3 to be driver-centric with little opportunity to determine how the distractive effects of the technology may be impacted by their integration with the wider sociotechnical system.

Legislation has been considered to be the predominant countermeasure to targeting driver distraction, but it has been critiqued for being a heavily individual focused method, seeking to fix the driver rather than support them (Tingvall et al., 2009; Young & Salmon, 2015). The AcciMap analysis in Chapter 5 highlighted, and analysed, the distinction in legislation between the drivers' use of handheld mobile phones and their use of other nomadic technologies. Notably, this highlighted the top-down influence that policy and legislation has on the lower level themes underneath it, which encompasses those who must abide by it. Yet, legislation tended to focus on the actions of the end user rather than the manufacturers, designers and regulators, particularly when it came to the use of handheld mobile phones. The legislation that targets the drivers' use of mobile phones is specifically focusing on the interaction between the end user and the mobile phone; it does not consider the interacting factors that other actors may have on this, such as device manufacturers. Furthermore, it may be creating the conditions for distraction from other technologies which are not targeted to the same degree in legislation and may therefore be deemed as safer alternatives (Chapter 5; Parnell et al., 2017a), when in reality they may not be (Redelmeier & Tibshirani, 1997; Strayer & Johnston, 2001; Horrey & Wickens, 2006).

The more abstract wording of the legislation that encapsulates all other potentially distracting tasks states that drivers 'should not be distracted' and should have 'proper control' over the vehicle. Yet, it does not state how this should be achieved or what 'proper control' means. This ambiguity was shown in Chapter 5 to have led to the development of technology in the vehicle that is not outlawed but that is also not fully tested for safe use while driving.

Thus, the research conducted in this book has suggested that current legislature is ineffective in dealing with driver distraction from nomadic and other in-vehicle devices. Focusing specifically on drivers' use of individual devices such as handheld mobile phones is not advised, not only does it lead to comparisons to other devices that are not illegalised but has also been ineffective in mitigating against the use of the handheld mobile phones. Increasing penalties within the UK law from £60 in 2007 to £200 and six penalty points in 2017 suggests that the measure has not been enough to stop drivers from using them. Indeed, the lack of references to legislation in the interview and driving studies conducted in this book suggest drivers do not widely consider the legality of their actions when engaging with technological devices. Instead it is proposed that more should be done by legislation to target the other levels above the end user in the RMF, such as the regulators, industrialists and resource providers. For example imposing tighter legislation on manufacturers at the design stage should seek to prevent distracting devices from being accessible to drivers. The actor map and AcciMap analysis in Chapter 5 presents the actors and interacting actions that can be targeted by legislation.

9.5.2 REGULATORS

Regulators in the road transport sociotechnical system surrounding driver distraction include those who regulate and enforce the laws set by the government, and the committees at the level above. As legislation is currently driver-focused, regulations are also focused on the driver. Surveillance and media campaigns have been rolled out in line with the policy on handheld phones to target the driver. Yet, as shown in Chapter 7, drivers continue to report their engagement with devices in a way that is little influenced by legislation or the fear of getting caught, but by factors relating to the task itself, their own perceptions, the road infrastructure and/or the context surrounding this. Therefore, the regulation of these other elements in the system is advised.

A recent measure to target mobile phone use in the UK has utilised the road infrastructure by developing road signs that are capable of determining if drivers are using their phones when they drive past. They aim to act as a deterrent, flashing up with a phone with a red line through it if mobile phone use is detected. These are currently under trial to determine if they make a difference, although there are still some challenges to overcome. For example they cannot differentiate between phone use by a passenger or the driver. Yet, such alternative solutions that utilise other systemic elements such as the road infrastructure are promising.

The regulation of technological devices, their implementation, design and necessity, would offer an alternative way of assessing the problem of managing the safe implementation of technological advancement within the vehicle. It was stated

Evolution of the PARRC Model   189

early on in this book that the issue with technological sources of distraction is their rapid development that often surpasses effective road safety tests (Stanton & Salmon, 2009; Leveson, 2011; Hettinger et al., 2015). This is especially true of devices whose initial purpose is not intended for the vehicle but they are used by drivers, such as mobile phones. Furthermore, developments in technology are often to suit the demands and desires of the end user (Hettinger et al., 2015), which in the vehicle can lead to competition between manufacturers to bring new features to the driver through novel interfaces (Ranney et al., 2000). Yet, priority should remain on safety.

Tighter regulation within the industry on the implementation and integration of new technologies into the vehicle should aim to amend this shift, re-centring the focus back to safety. This will of course be difficult as manufacturers begin to introduce increasing levels of automation into the vehicle, which will lead the driver to be less engaged in the driving task and more free to engage with other tasks (Jamson et al., 2013; Stanton, 2015). Yet, this will provide an even greater need for regulations within the industry (Fagnant & Kockelman, 2015). This is an issue that EURO NCAP is aiming to tackle in their outlined plans for the implementation of automation from the year 2020 to 2015 (EURO NCAP, 2017).

9.5.3   INDUSTRIALISTS

Manufacturers should have a responsibility to monitor and prevent unsafe use at the device level, rather than at the end-user level which is more difficult (WHO, 2011). The development of driver-monitoring systems will be important to the implementation of increasing levels of automation in the vehicle (EURO NCAP, 2017). The implementation of these technologies that allow for driver-state monitoring and warning systems will enable the developments in technology to provide a positive impact on driver safety. Yet, the importance of accuracy and driver acceptance are paramount to their success (Dong et al., 2011). Apple has recently taken a step towards this with the implementation of their 'do not disturb the driver mode' to their mobile devices that, as discussed in Chapter 7, suggests they are aware of their role in driver distraction from mobile phones and are taking some responsibility for the mitigation of the issue. The high proportion of drivers who stated they would glance at a text message when they receive it, would be prevented from acknowledging the text when this mode is selected. Yet, the other features that are available on such smart phones may suggest this move to be quite hypocritical as developments in other technologies such as voice-activated communication systems and navigation features were reported to be used by a number of participants when driving in the interview and driving studies. For example Apple CarPlay® allows the in-vehicle interface to replicate the applications and functionalities of the phone. The multifunctionality of the mobile phone was found to be a detrimental to the drivers' engagement with the mobile phone device in the interview study conducted in Chapter 6. Particular concern was raised on the reliance on mobile phones for navigational information that has led the driver to place their phone in a visible and easy-to-reach location while driving, which has been facilitated by the development of phone fixture devices within the market (Chapter 5). A more co-ordinated approach to distraction

190 Driver Distraction

is required among industries. For example improvements to specific satnav devices that utilise the facilities drivers have come to rely on their phones for, such as the Google Maps application, could prevent drivers from having their phone on display while they are driving, limiting its multifunctionality.

9.5.4 Resource Providers

As was highlighted in Chapter 5, companies that provide drivers with technology need to be aware of how they may be diverting the drivers' attention away from the road. This level of the system is closely intertwined with the end users' interaction with the device, therefore it should seek to increase the efficiency and safety of the drivers' interaction with technology to reduce the attentional resources that are diverted away from the road. This may be through applying Human Factors design principals to interface design, or by freezing out functionality while the vehicle in is motion. It was highlighted when applying the PARRC model to the case study of Victoria McClure that the incident could have been avoided if the functionality to enter an address while in motion, which was a task that took longer than the 12 s guidelines recommended by NHTSA (2013) (Harvey et al., 2011c) was frozen out by the device developers, as it is by some other manufacturers. As suggested previously, this difference may reflect the different design cultures of manufacturers and their approach to distraction regulation (NHTSA, 2013). Restrictive interfaces, characteristic of Japanese manufacturers, limit the drivers' potential to self-regulate their engagement with devices which can stop them from placing themselves at risk of distraction-related events. Yet, this does not always eliminate the issue as adaption to overcome restrictions can occur, with drivers seeking ways around the limitations imposed on them. For example putting the handbrake on while in motion to allow interaction with the interface while driving. Such incidents are evident within the sociotechnical systems approach as emerging from the adaption of behaviour incrementally over time with the adjustment of normal behaviour until it reaches a point where it is a departure from the previously accepted norms and safe operations (Dekker, 2011).

The complexity of systems makes unanticipated circumstances difficult to realise which is why looking for cause/effect relationships between actors is reductionist (Dekker, 2011). This is underpinned in the 'drift into failure' model that is based within the sociotechnical systems approach (Dekker, 2011). This details how behaviour adapts slowly overtime from normal, safe practises that are subject to efficiency and financial pressures which can lead to adaptive shifts in behaviour towards a boundary of functionally acceptable behaviour. At this boundary, safety is deemed to be compromised. Salmon, McClure & Stanton (2012) highlight the importance of this process, termed 'decrementalism' (Dekker, 2011), to the developments in in-vehicle technology in the vehicle. Technological devices such as in-vehicle interfaces and mobile phones have become increasingly more advanced since their initial implementation. The incremental advancements in these systems means that guidelines and standards have a challenge in keeping pace. They can also have unanticipated effects which is why testing within limited parameters, with objective measures outside of the real-world context are not always adequate.

Evolution of the PARRC Model

The 'task' was found to be one of the key systemic themes in the inductive thematic framework developed from the interviews in Chapter 6 and validated in Chapter 7. Task subthemes included the 'ability to complete', 'complexity', 'desirability', 'duration', 'interaction' and 'engagement regulation'. These themes and the drivers' perspective could be used to inform the design of future interfaces to lower the drivers' likelihood of engaging at inappropriate times.

Limiting task duration and improving the interface of devices are ways in which these actors can mitigate against the distractive effects. Yet, the desire that drivers have to engage with the device has been less considered. Thus, alongside the design of the task, the desire that manufacturers are placing with the drivers to engage with tasks that compete with the main driving task, and therefore the maintenance of safety, needs to be more widely considered. This requires looking at how the development of technologies and their use are impacting on the end user. The possibilities for alternative, safety-conscious ways of allowing the drivers to interact with their phone may be another way to approach the issue, for example driver-monitoring technology (Dong et al., 2011; EURO NCAP, 2017). The use of head-up displays and the optimisation of information to meet the demands of the environment are possibilities that have been considered (e.g. Liu, 2003).

It is also important to consider the role of culture within interface design. A study conducted by Young, Rudin-Brown et al. (2012) contrasted drivers from Australia with drivers from China and found cultural differences in the design preferences and interaction requirements. The Chinese cohort were more focused on the aesthetic of the in-vehicle infotainment systems (IVIS) than its functionality and usability than the Australians. For the Chinese, the high-end, modern interfaces were preferred as they suggested a sense of high status, whereas the Australians were more concerned with safety (Young, Rudin-Brown et al., 2012). The Chinese were also more likely to prefer flat information hierarchies in interfaces with more freedom to explore and less structure. While Australians were more likely to want to customise and configure the IVIS design. There were also some similarities between the two cultures, such as both preferred symbols over abbreviations or words on the interface. Yet, the Chinese had obvious difficulties in comprehending the English abbreviations in contrast to the Australians. Thus, the cultural references are important to consider when researching the drivers' perceptions on in-vehicle tasks and their ability to complete them.

9.5.5 End Users

The focus of this book has aimed to shift the responsibility for distraction away from solely targeting the individual and towards accounting for the surrounding factors that influence the drivers' interactions with distracting technologies, that is those at the hierarchical levels of the system above. This does not, however, suggest that the driver may not hold some responsibility. Indeed, the 'driver' was one of the systemic themes identified in the thematic framework in the interview study. Yet, it is stated that accountability for other actors that interact and guide driver behaviour in the emergence of distraction should be recognised in order to provide effective mitigation strategies.

The interviews revealed that the drivers' attitude was a key influencer of their likelihood of engaging with the technological tasks. This aligns with other research that has applied the Theory of Planned Behaviour to the domain (e.g. Chen & Donmez, 2016). Reviewing the systemic actors showed that legislation, media campaigns, road safety charities and training providers were potential sources through which to change the attitudes of drivers. Their spread across the hierarchy from the top and down, showed the impact of legislation on shaping attitude through the regulators and education providers. The potential to change the perceptions of the driver with the use of these higher hierarchical levels was identified. While this suggests that legislation can impact on the driver, the support of the hierarchical levels in-between the two systemic actors is required. For example improved HMI developed by manufacturers can minimise the attentional resources that are required to interact with the technological devices. Workload managers that disable certain features that demand resources that are required in the driving task also prohibit the driver from diverting their attention from the road environment to interact with secondary tasks.

Chapter 7 assessed the effect of age and gender on the drivers' discussions on their likelihood of engaging with secondary tasks ratings. Younger drivers rated themselves to be more likely to engage with technologies than older drivers. This effect was linked to the younger drivers' perceived ease of interacting in contrast to the older drivers (Parnell et al., 2017b). This supports previous findings that suggest younger drivers are more confident in interacting with secondary tasks while driving (Alm & Nilsson, 1995; Reed & Green, 1999). Yet, as older drivers are less likely to interact with the technology (McEvoy et al., 2006), different countermeasures used to target drivers of different age groups are warranted. Many countries do have increased penalties for young or novice drivers. Recent amendments to the UK legislation now mean that drivers who have had their licence less than 2 years could risk losing their licence if they are caught using their phone. Yet, as this book strives to demonstrate alternative countermeasures to individual focused legislation, more innovative countermeasures towards the younger age group are required. Targeted media campaigns and increased awareness throughout driver education may be one such method. Indeed, the driving test in the UK was recently updated to include a section where the driver must show they are able to safely follow directions from a satnav. The impact that this has on technology use and distraction statistics will be of interest in the future.

9.5.6 Equipment and Environment

Recommendations at the lower levels in the environment are hard to implement as they relate to the contextual factors surrounding the behaviour. The road infrastructure was found to have an influence over the drivers' intention to engage with technological tasks in the interview study that assessed the seven road types identified by Walker et al. (2013). This indicated, and was supported by findings in the driving study, that drivers were able to strategically plan their engagement with the technological tasks to some degree. Notably, drivers stated they would be most likely to engage when stopped at a junction and waiting at a red light, than if they were driving through the junction (Chapter 6). Participants in the interview study stated that

Evolution of the PARRC Model

they did not deem themselves to be 'driving' when they were stopped at lights and therefore thought it provided a good opportunity to engage. The strategic engagement with the tasks in line with the demands of the infrastructure suggest that drivers are aware of the risks of engaging in secondary tasks relative to the environment. This may be evidence to suggest that allowing drivers to be free to choose when to engage with technologies, as is the approach used in Sweden, may be a better method than the outright banning of devices. Before adoption of the Swedish approach, strong research evidence is required. A suggestion by Tivesten and Dozza (2015) may provide an alternative, with specific lanes of the road permitting technology engagement in regulated areas that can allow drivers to strategically interact with the tasks when necessary. Specifically informing road users in these areas that they may interact with their devices may facilitate safety enhancement measures that may counter the adverse effects of engaging with technological distractions. For example, legislation with variable reinforcement of the drivers use of technologies that may allow them to interact when they are stopped at traffic lights for a period of time but prohibits them from engaging while they are driving may minimise the time drivers spend attempting to engage while in motion, encouraging them to postpone it to a point where they are permitted. This may have the potential to limit distraction engagement within safety critical situations. Further research would be required to assess how these may be integrated with the current road transport system. Furthermore, this will have to adapt with the developments of automation within the driving task that will permit the driver to be less engaged with the driving task for longer durations of time which may allow them to be more flexible with their interactions with technology.

Assessment of the recommendations that could be made to the 'context' system theme (from the thematic framework in Chapter 6, Figure 6.6), identified that there were no actors that influenced the theme at the higher levels of system. The main influencers over context were much more prevalent in the lower levels of the system, emerging from the environment and the user themselves. Research centres and universities were, however, presented as an actor further up in the system who should seek to identify the effects of different contexts, situations and conditions that impact the emergence of distraction from the system. The ability to research distraction in the natural driving environment is difficult, with experimental noise and safety issues (Carsten et al., 2013). Evidence in Chapter 8 shows that, with a full-car simulator, drivers' decisions to engage in secondary tasks can be effectively assessed.

9.6 CONCLUSION

This chapter has reflected on the PARRC model of distraction (Parnell et al., 2016), which has underpinned the theoretical, methodological and practical developments of this book. The evolutionary trajectory of the model across the methodologies applied has been documented to show the benefits of applying the model as well as validating it through the process of triangulation. The insights that have been obtained have informed a novel, systems-based definition of driver distraction.

The reason for developing a novel model of distraction that highlights the importance of systemic factors was to enable the potential for alternative countermeasures to prevent the behaviour. Previous attempts to mitigate driver distraction, particularly in response to rapidly developing technology, have relied on the use of legislation that targets the individual driver. Taking a systems view of the issue suggests that focusing solely on the individual is not effective. Instead the interaction of a host of other actors within the system needs to be addressed. The evolutionary processes of the PARRC model has found key factors of distraction that are influenced by, and interact with, key systemic actors. These actors offer the potential to provide novel countermeasures to target technological distractions in the vehicle.

The risk management framework details the hierarchical levels of a sociotechnical system and can be used to inform on the actors comprising a system (Rasmussen, 1997). Assessment of the systemic actors that were identified from the application of the PARRC model throughout the work conduction in this book were reviewed to provide recommendations for countermeasures to technological distractions. The distribution of these actors across the hierarchy has been presented. This has shown how the research methods that have been applied can inform recommendations in order to minimise the research-practise gap in the field that was identified in Chapter 3 of this book.

10 Conclusions

10.1 INTRODUCTION

This concluding chapter will provide an overview of the work that has been presented in this book and how it has resolved to fulfil the aim and objective set out in the introductory chapter. It will then discuss the implications for theoretical, methodological and practical developments, as well as examining the potential for future work as a direct outcome of the work presented in this book.

10.2 SUMMARY OF FINDINGS

The following section will present a summary of the findings evident from this book in relation to each of the key objectives. These objectives are as follows:

1. *Review the current methodologies that are used to study driver distraction and seek out methods that can assist in the application of the sociotechnical systems theory to the study of driver distraction.*

   The first objective aimed to review the methodologies that have been, or could be, applied to the study of driver distraction and determine which may be used to study driver distraction from the sociotechnical system perspective. It was understood, in Chapter 3, that there are multiple methods that can be used to study the issue of driver distraction. Yet, caution was heeded to first assess the outcomes that were required, before establishing the methods that could be used to achieve them.

   Traditional dichotomies between qualitative and quantitative methods, as well as subjective and objective approaches, were identified in Chapter 3 (e.g. Ercikan & Roth, 2006; Denzin & Lincoln, 2011). The inter-relations between these dichotomies were realised with respect to the domain of driver distraction. Objective quantitative measures, which are common in the research field, measure the impact that distractive tasks have on the primary driving task to enable a quantifiable output; such as the length of time drivers looked away from the road (e.g. Green, 1999; Harvey & Stanton, 2013). Objective, qualitative measures can be used to determine the magnitude of the issue (e.g. Lam, 2002; Sullman, 2012). For example, an analysis of crash reports by Beanland et al. (2013) determined the role and nature of driver distraction and inattention in a large sample of incidents, which led to the assessment that 70% of distractions were voluntarily engaged by drivers. It is these objective measures that have highlighted the issues that secondary tasks, such as those posed by technologies within the vehicle, have on the driving task and the impact that the issue has on road safety. Conversely, subjective measures aim to gain the drivers' perception of their behaviour, their views on engaging with secondary tasks while driving and how they would go about engaging with secondary tasks (e.g. Lerner, 2005; White et al., 2004). These methods have, however, been less frequently applied to the driver distraction domain. Subjective insights can be gained by asking drivers to quantitatively rate their likelihood of performing certain tasks while driving, or engaging them in more open-ended, qualitative, discussions on their motivations and beliefs. It was proposed in Chapter 3 that subjective methods should not be disregarded in the driver distraction domain. Instead, they should be used to enable a deeper understanding of why drivers engage with distracting tasks, rather than what happens when they do (as this has been understood from objective methods). This advice was heeded with the application of subjective, qualitative methods throughout Chapters 6–8 of this book.

Furthermore, in line with the movement within the Human Factors domain away from 'human error' and towards a complex systems approach to accidents (Stanton & Harvey, 2017), it has been suggested that focusing on the driver as the main causal factor contributing to distraction-related events has limited effects (e.g. Stanton & Salmon, 2009; Young & Salmon, 2012). Instead, it is important to understand the impact that other elements may have on the drivers' interaction with distracting tasks. This suggests the need for the use of methods that are able to capture the influence of these 'other elements' with methodologies that are underpinned by systems thinking (Young & Salmon, 2012). Hettinger et al. (2015) stated that 'traditional experimental paradigms may not be equipped to adequately examine or explain complex system performance' (p. 601).

A research-practise gap, identified in other accident analysis domains (Underwood & Waterson, 2013), was also found in driver distraction (see Chapter 3). This highlighted a movement towards the development of systems research into road safety, but a lack of applications of this approach in the management of road safety behaviours, including driver distraction. An avenue for research was evident in determining why driver distraction is occurring, as influenced by the wider systems factors, and their contribution to the adverse events that are caused by the drivers' interaction with distracting technologies. This has formed the basis for the work in this book; applied both systems-based methodologies such as the AcciMap analysis in Chapter 5 and subjective methods including the semi-structured interviews in Chapters 6 and 7 as well as the verbal protocol analysis in Chapter 8. Each of these methods was applied to the study of driver distraction from in-vehicle technology to enable insights into the systemic causes of the behaviour. They have helped to reveal the interactions between systemic actors and their responsibility for distraction-related events, which has enabled a shift in focus away from the driver towards other possible sources of mitigation techniques.

2. *Develop a framework through which to study driver distraction from a sociotechnical systems approach.*

Conclusions

The need to assess driver distraction from a sociotechnical systems approach was realised in Chapter 2. The progression of systemic methods (e.g. Rasmussen, 1997; Shappell & Weigmann, 2012; Leveson, 2004; Stanton & Harvey, 2017) and the possibilities they held to reveal why distraction occurs and how alternative countermeasures may be developed to effectively mitigate against it were established in Chapter 3. It was therefore required that any causal factors linked to distraction should be able to inform, not just on the drivers' role in distraction but also on the interconnections to the wider sociotechnical system. It was evident in the literature that there was limited research into the casual factors involved in distraction and how it results in error (Stanton & Salmon, 2009; Young & Salmon, 2012). Chapter 4 details the procedures, findings and conclusions from the application of grounded theory methodology to the literature on driver distraction from in-vehicle technology to determine the key mechanism that lead to distraction. Grounded theory led to the development of a novel model which comprised the key factors from the literature on driver distraction from in-vehicle technology. Five key factors were identified: 'goal priority', 'adapt to demands', 'resource constraints', 'behavioural regulation' and 'goal conflict'. These are defined with examples in Chapter 4 (Section 4.3.1). The interactions between each of these factors, cited within the literature, were used to combine the factors into the PARRC model of distraction (Parnell et al., 2016).

The ability of the PARRC model to capture the influence of wider systemic factors was first realised in its application to the case study of Victoria McClure (Chapter 3; Parnell et al., 2016). The AcciMap analysis completed in Chapter 4 built on this, showing the relevance of the factors across all levels of the risk management framework (RMF) (Rasmussen, 1997), as well as the additional levels that were established in the international and national committees that inform governments (Parnell et al., 2017a). Not only did the AcciMap analysis present the array of systemic factors that are involved in driver distraction from technological devices, but it also revealed the distinction that is currently evident in legislation, between mobile phones and all other nomadic technologies (Parnell et al., 2017a). This highlighted the impact that legislation, at the top of the system, had on all of the systemic interactions across the levels below it through the PARRC factors of distraction.

Further exploration of the drivers' views on their use of technological devices while driving showed support for the factors of distraction. The inductively generated thematic framework of the drivers' reasons why they may be more or less likely to engage with technology was aggregated into four high-level systemic themes: the wider context, the task, the infrastructure and the driver. Evidence for these levels was found in Chapter 8 when drivers were asked to state their reasons for engaging, or not engaging, with different technological devices while driving on the roads and in a simulator. This validated the original framework, while also providing the addition of another theme: other road users.

The insights gained on the actors within the sociotechnical system has shown how they influence the factors of distraction and therefore how they are involved in the events leading to distraction-related accidents. This has ramifications for countermeasures that can target the wider sociotechnical system surrounding distraction, rather than just the individual driver. These were summarised in the review of the development and application of the PARRC model in Chapter 9.

3. *Assess the current countermeasures that are used to tackle driver distraction from technological devices and determine the potential for novel countermeasures.*

The disparity in the research field between the 'old' view of accident causation that resolves to target the individual and the 'new' systems view that considers the wider interactions of the driver with the sociotechnological system in which the behaviour occurs (Reason, 2000; Dekker, 2002) is discussed in Chapter 2. Legislation that attempts to ban drivers from becoming distracted, and penalising them if they do, is predominant in traditional countermeasure to driver distraction, and the road incidents more generally (Salmon et al., 2012). Critique of this method is proposed by those promoting the systems approach, often by highlighting the limited view of the traditional (individual) methods (Young & Salmon, 2015; Hignett et al., 2015) as well as its ability to only be employed with hindsight, prosecuting drivers only once they been found to be distracted (Kircher & Ahlström, 2016).

The AcciMap analysis in Chapter 5 highlighted how legislation targeting mobile phones was unable to prevent distractions from other technologies. It was suggested that it may actually assist in creating the conditions for drivers to engage with other technologies as it highlights the comparative risk of mobile phone use (Parnell et al., 2017a). As an alternative, the actor map in Chapter 5 presents the array of actors in the sociotechnical system surrounding driver distraction from in-vehicle technology that may be the subject of systemic countermeasures. The application of the PARRC model to the AcciMap suggests how each of the actors at the various levels of the system may support, or conversely leave unsupported, the causal factors of distraction.

Utilisation of the RMF (Rasmussen, 1997) to highlight the recommendations for possible countermeasures across the hierarchical levels of the system was first presented in Chapter 5 in response to the AcciMap analysis (which was extended to include national and international committees). It was further explored in Chapter 5, where the systems themes that are identified at the systemic levels of the thematic framework were discussed with respect to the hierarchy of the RMF. The RMF was further applied in Chapter 9 to gather the recommendations that had been gleaned from the development, application and validation of the PARRC model across the chapters of this book. The benefit of providing recommendations across the RMF is the graphical representation of the system. The recommendations made from the sociotechnical system approach highlight the interacting nature of the hierarchical levels in their encouragement of the safe performance of the system. Conversely, traditional methods have tended only

Conclusions 199

to focus on the legislation at the top of the system and the end user at the bottom, with little reference to the levels in-between.

Notably, the recommendations presented have highlighted the social responsibility that the system has in influencing the drivers' attitudes and shaping their behaviour through media campaigns, educational providers and road safety campaigns. It also suggests the relevance of infrastructure design, environmental conditions and task context on manipulating the complexity, motivation and consequences of interacting with technological tasks that need to be understood. Yet, arguably, the most appropriate countermeasure may be realised by targeting manufacturers. Whilst current legislation aims to prevent the driver from interacting with technology, it does not strive to prevent the technology from interacting with the driver. Hettinger et al. (2015) state that the advancement in technologies are shaping human needs in such a way that the maintenance of safe systems now falls to those who caution the risk of the advancements that technologies are having within systems. Yet, it is the developers of technologies who facilitate an individuals' access to information, entertainment and communication devices (Walker et al., 2001; Ranney et al., 2000). The applicability of this access across contexts needs to be realised. It is noted that error is considered to arise from systems that permit and, in some cases, encourage behaviours that turn out to be erroneous, with the benefit of hindsight (Reason, 1990). Manufacturers, therefore, have a responsibility to maintain safety over the facilities they provide. The NHTSA (2013) guidelines have outlined standards that manufacturers must adhere to in their design and testing of in-vehicle interfaces. Yet, the testing procedures are also driver focused, with experimental recommendations given for simulator studies and occlusion testing which do not account for the wider system implications (Parnell et al., 2017b) or the adaptive constraints of the sociotechnical system (Dekker, 2011). Safety should be defined by the context of use (Harvey et al., 2012), consequently manufacturers of nomadic devices should be aware of the different contexts within which their technologies are used, such as in the safety critical context within the vehicle.

The technology manufacturer Apple has recently taken steps towards this with the development of a 'do-not disturb while driving' mode on their mobile phone devices which prevents their users from interacting with the phone and receiving notifications when it detects they are in a moving vehicle.

10.3 FUTURE WORK

The work presented in this book has developed an understanding of driver distraction from in-vehicle technology, the factors that relate to the emergence of distraction and the potential for applying a sociotechnical view of the problem to provide alternative countermeasures to those suggested by the 'traditional' approach (Reason, 2000). The implications of this work to inform future theoretical, methodological and practical development are discussed.

10.3.1 Theoretical Implications

10.3.1.1 Sociotechnical Systems Theory and Definition of Driver Distraction

This book has aimed to review and develop the potential for a sociotechnical systems approach to driver distraction. The development of the PARRC model has assisted this process and shown the potential for a sociotechnical systems model of driver distraction from technological devices. It is anticipated that the insights gleaned from this research will inform the development of the sociotechnical systems theory in the driver distraction domain as well as the road transport domain more generally. Whilst it was highlighted in Chapter 2 of this book that there is no universal definition of driver distraction, nor a widely agreed theory from which to study it, it is hoped that the work presented here shows the importance of reviewing the wider sociotechnical system in future driver distraction research. The PARRC model, and its associated definition of distraction that is stated in Chapter 8, provide a novel addition to the sociotechnical systems approach to driver distraction that should be further explored and applied by researchers in the future.

10.3.1.2 Further Exploration of the PARRC Model

While the work in this book has contrasted the development of the PARRC model from the literature to its development from the drivers' perspective, further work should seek to continue to explore the model with the creative application of methodological approaches that facilitate the sociotechnical systems perspective. The development of the model in Chapter 3 was exploratory to determine how a systems model may be able to explain the sociotechnical approach to driver distraction and determine the causal factors of distraction stated in the literature. The developments that have been presented across the chapters of this book suggest that there is scope in the literature for a sociotechnical systems view of driver distraction.

The evolution of the PARRC model in its application to the drivers' reports in the interviews, and when driving, suggest there may be some gaps in the literature from which the PARRC model evolved. This includes the connection between the factors 'resource constraints' and 'goal priority'. Further research is required to explore these gaps.

10.3.1.3 Cultural Differences

The work conducted in this book has been predominantly informed by knowledge of the UK road transport system. The legislation of the UK informed the AcciMap analysis conducted in Chapter 4 and UK drivers were used in the interview and driving studies that were conducted across Chapters 5–7. Whilst the UK legislation on mobile phone use and other technology use in the vehicle is similar to that established globally, the cultural norms and road transport context is likely to differ. The findings of Young, Rudin-Brown et al. (2012) showed that drivers from difference cultural backgrounds value different aspects of interface design and this affected their interaction with it. Furthermore, different cultures are influenced by different manufacturer guidelines (e.g. Japan Automobile Manufacturers Association, JAMA, 2004; Commission of the European Communities, 2008; Alliance of Automobile

Conclusions 201

Manufacturers, 2006). Each of these are considered within the NHTSA (2013) guidelines but companies have already adopted different approaches to interface design principles. For example Japanese companies favour restricted access to complex functions, whereas European companies favour offering the driver more choice over when they can interact with tasks whilst driving.

The sociotechnical systems approach incorporates the complexity of the social-organisational and technical factors that influence behaviour and the context within which it occurs. Other countries will have different norms surrounding technology use and driving behaviours. They may also have alternative ways of providing law enforcement and influencing cultural views towards distracted driving. Therefore, it would be of interest to contrast the findings conducted within the context of the UK to other countries. The AcciMap analysis may provide alternative interdependent relationships between factors in the RMF and an inductive thematic analysis may reveal different concepts of interest that inform the drivers' decision to engage with technological devices while driving. This may assist in furthering our understanding of the role of culture and context to sociotechnical systems and its impact on the issue of driver distraction.

10.3.1.4   Context and Road Environment

It was highlighted in the recommendation section of Chapter 5 (Section 5.4.2) and Chapter 8 (Section 8.7.1) that the context within which driver distraction occurs is highly involved in its emergence from the sociotechnical viewpoint. Yet, its relationship to distraction-related incidents and the ability to prevent them are little understood. The role of research centres was revealed to be a key actor that may be able to identify this relationship and assist in the development of potential countermeasures. The work completed in this book sought to understand how different road types may be perceived by drivers and how this may influence their decision to engage with different technological devices. While road type was not found to greatly influence the decision to engage as much as task type, a distinction between the drivers' perceptions when they are stopped in traffic (e.g. at a red light), compared to when they are driving through a junction, should be more closely explored. Furthermore, the driving study detailed in Chapter 7 showed that drivers were significantly less likely to engage with technological tasks when driving on roundabouts than on A roads or motorways. Yet, behaviour at other junctions was not explored. This could be explored further, building on the methodologies developed within this book, for example verbal protocol and naturalistic decision making.

10.3.1.5   Automated Driving

The focus of this book has been on manual vehicles that require the driver to play an active role in vehicle control, which is why distraction from the driving task can lead to adverse safety implications. Yet, the developments in vehicle automation are occurring at a rapid pace, with vehicle manufacturers committed to the 'driverless car' concept (see other books listed in this series). While the development of automated vehicles offers the opportunity for enhanced safety in the road transport domain (EURO NCAP, 2017), the transition towards automated vehicles requires careful consideration from a legal and research perspective (Brookhuis et al., 2001; Fagnant & Kockelman, 2015).

Whilst technological capability of the vehicle has advanced to produce automated vehicles, Human Factors research is important to enable the driver to successfully interact with the vehicle as well as to be aware of its capabilities and limitations (Merat et al., 2012; Merat et al., 2014; Banks & Stanton, 2014a).

The implementation of automation within systems has been found to put the driver 'out-of-the-control-loop', thereby reducing their situational awareness as they adopt the role of monitor and supervisor rather than active controller (Stanton & Marsden, 1996; Byrne & Parasuraman, 1996; Banks & Stanton, 2014a). The development and implementation of adaptive cruise control (ACC) in the 1990s to early 2000s identified that while the automated features of ACC may increase safety in the control of vehicle (Stanton & Young, 2005), it may have implications on the drivers' situational awareness and their ability to safely regain control, especially when engaging with secondary tasks (Rudin-Brown et al., 2003; Ma & Kaber, 2005).

Driving automation is advancing across the levels of automation set out by the Society of Automotive Engineers (SAE, 2016) from partial automation (Level 2) where the driver plays a large role in the dynamic driving task with the automation assisting with steering and accelerating/decelerating, to Levels 3 and 4 where the driver plays a monitoring role, only responding to the vehicle as and when required. This progression will free up additional attentional capacity that will allow the driver to engage with 'other' tasks in the vehicle. Yet, if drivers are able to engage with other activities while driving but are also still required to engage with the vehicle where required, then the transition of attention will be important to the safety and success of increasing levels of automation within vehicles (Jamson et al., 2013; Merat et al., 2012; Merat et al., 2014; Banks & Stanton, 2016; Young & Stanton, 2002, 2004). As automation is likely to reduce workload and allow the driver to disengage from the driving task, the need to re-engage and resume control when the vehicle and environment demand it can have adverse safety effects, as the driver tries to adjust their workload and become fully immersed in the driving environment (Rudin-Brown & Parker, 2004; Merat et al., 2012; Stanton et al., 1997). Indeed, many manufacturers are pre-empting the move towards automated driving and the drivers' desire to engage with secondary task by providing increased functionality within partially automated vehicles. The ability to allow drivers to remain in the loop when in automated vehicles while also allowing them to benefit from the potential that automated vehicles have to facilitate engagement in other secondary tasks is a requirement of driver attention research going forward (Merat et al., 2014). Additionally, there is also discussion on whether sources of distraction may actually become facilitators of workload transition in automated vehicles (Miller et al., 2015; Young & Stanton, 2002, 2004).

Theories of driver attention and distraction are therefore still relevant as the automotive domain progresses. Although, their revision and re-application to the process through which the driver interacts with vehicle automation needs to be addressed. The challenges that are faced with respect to the integration and application of automated vehicles are particularly pertinent to the sociotechnical systems approach (Fraedrich et al., 2015; Banks et al., 2018). The integration of automation will cause a fundamental change within the sociotechnical system that will mean taking a holistic approach to a problem will be important, yet challenging (Fraedrich

Conclusions

et al., 2015). The application of the principles and actors of the sociotechnical system that have been highlighted as relevant to driver distraction will be of importance when reviewing automated vehicle integration: international and national committees (SAE, 2016; NHTSA, 2017), legislation, regulation (Department for Transport, 2015c) and manufacturers (EURO NCAP, 2017). Furthermore, the key concepts of the PARRC model are still relevant to assessing the goals of drivers and automation. The implementation of automation may reduce the conflict that secondary tasks have on the driving task as automation will free the drivers' attentional resources to prioritise other activities. It will be interesting to determine how drivers adapt to the new demands of the driving task and automation as well as how they regulate their behaviour in response to changes in conflicting goals when the vehicle requires them to resume control.

10.3.1.6 Applications to Other Domains

As mentioned in the introduction (Chapter 1), the focus of this research has been on drivers of standard passenger vehicles and their everyday use, yet driver distraction is a phenomenon that is prevalent across multiple other transport domains, including rail (Naweed, 2013), aviation (Loukopoulos et al., 2001), long-haul truck drivers (Hanowski et al., 2005), bus drivers (Salmon et al., 2011) as well as pedestrians (Schwebel et al., 2012) and cyclists (de Waard et al., 2010). Furthermore, technological advancements have been found to provide distractions in other domains including the classroom (Campbell, 2006), healthcare (Papadakos, 2013) and the workplace (Korpinena & Pääkkçnen, 2012). The insights that have been gleaned from the research conducted in this book that have applied sociotechnical systems thinking to distraction and the associated methodologies that have been used may offer similar benefits to these other domains. There is also the possibility of applying the PARRC model of distraction to these other domains to explore how it may be able to explain the behaviour and develop insights into the wider sociotechnical system within which the distractive behaviour occurs. Comparisons of the model across other transport domains could provide interesting differences in how distraction emerges across domains and therefore how they may differentially, or commonly, mitigate against it.

10.3.2 Methodological Implications

10.3.2.1 Qualitative Research

This book has shown the insights that can be obtained from the application of qualitative methods to the study of driver distraction from the drivers' perspective. This has shown that subjective methods may be able to gain insights that cannot be reviewed using objective methods. Obvious trade-offs have been made when using the qualitative methodologies, between sample size and rich data analysis. The utilisation of an online survey to compliment the findings from the interview study suggested that the interview participants were representative of the larger survey sample. Further work should, however, seek to further explore the findings in relation to a wider subset of drivers across demographic groups. Now that the thematic framework has been developed and validated with its application to simulated and on-road driving, the possibility of applying it to the verbal protocols and interviews of drivers in a more deductive process may enable data collection with much larger sample sizes than those used for the initial inductive thematic analysis.

The use of verbal protocols to assess the drivers' decision-making processes was a novel application of the methodology in this domain. It showed many benefits to identifying the factors that influence the drivers' intention to engage, in vivo. There were some mixed findings on how this may have affected drivers' performance, which require further investigation before conclusions can be drawn. Further exploration of the effects of verbal protocols to assess the drivers' decision-making process should be conducted with a greater range of driver performance metrics, including lateral measurements, steering wheel angle, and brake and throttle inputs.

10.3.2.2 Experimental Setting

Assessment of the drivers' decision to engage with technological tasks while driving found that their reasoning was strongly correlated between the on-road and the driving simulator studies. This suggests the benefit of utilising driving simulators to assess the drivers' intention to engage, as well as their actual engagement, within a controlled and safe environment. The simulation route used was, however, the product of a substantial body of work to recreate the same route as that which was driven in the real world, with the inclusion of local landmarks and realistic driving conditions. The strong positive phi correlation found in Chapter 7 validates the use of this simulated route for future research that can contrast realistic driving around Southampton with a simulated version of the same route. This presents the opportunity for further Human Factors research that seeks to compare experimental environments and their effect on the driver. It also provides validation of the simulation for other studies that are seeking to provide realistic driving environments. This may be of particular use to the increasing developments in automated driving research.

Further work, in response to the findings given in Chapter 7, should assess how the drivers' decisions may alter if they were able to interact with the technological tasks in the simulator environment. Assessment of their driving performance in contrast to their decision making and verbal reports could be conducted in this safe environment with strong evidence to suggest its relative validity to findings on the road.

10.3.2.3 Eye Tracking

Driver distraction research is often, and advisably, supported with the use of eye-tracking methodologies that can determine the drivers' visual behaviour alongside their behaviour in the driving task and the secondary tasks (e.g., Metz et al., 2011; Harvey & Stanton, 2013). As identified in the evaluation (Section 9.3), the addition of eye-tracking data to the driving study in Chapter 7 would have added greater insights into the drivers' decision-making process and the areas of interest that contributed to the decision. Future work should seek to build on the findings of Metz et al. (2011) on the visual behaviour drivers perform when deciding to engage with a secondary task and how this relates to the qualitative insights gleaned from the verbal protocol methodology. Comparisons between the drivers' glance behaviour when verbalising their intention to engage with secondary tasks could be made to their glance behaviour when they are actually asked to complete distractive tasks

Conclusions 205

to assess the validity of the verbal protocol methodology. This requires advanced eye-tracking equipment that is able to capture the location of the drivers' fixation in relation to their head position across their entire field of view.

10.3.3 Practical Implications

10.3.3.1 Recommendations to Practise for Alternative Countermeasures

This book has raised concerns regarding the use of legislation to mitigate against driver distraction in the UK, as well as other countries. It has also shown the potential for current legislation on the use of mobile phones by drivers has to 'create the conditions' for distraction from other technological devices. The backdrop of rapid technological advancement and its integration with everyday life is making the situation worse rather than better. The sociotechnical approach that has been taken within this book has sought to present the hierarchical levels of the system within which driver distraction emerges, and the actors within the system that influence its emergence. This has highlighted the role and responsibility that the levels of the hierarchy have over the development of distraction, and therefore hold the potential to prevent distraction. The importance of efficient legislation has been suggested and the extension of the RMF, originally developed by Rasmussen (1997), to include the national and international committees, who lobby the legislation set by governments. The involvement of these extra levels within the sociotechnical system and the development of countermeasures to distraction should be reviewed in future applications of the RMF to this issue.

Future efforts to counter driver distraction should seek to investigate the actors involved in this sociotechnical review, particularly those emerging from the AcciMap analysis in Chapter 4, for example at the 'regulators', 'industrialists' and 'resource providers' levels of the sociotechnical system. This has placed particular emphasis on the manufacturers of devices who need to fully test the functionality and applicability of their device(s) with relation to the wider sociotechnical system to assess how they are integrated and adapted to by the system as a whole. Steps have been taken recently by the phone manufacturer Apple, who have developed a 'do not disturb while driving' mode that encourages drivers to limit phone functionality while driving. Yet, further integration across industries is suggested to tackle other technologies that are both brought into the vehicle by the user and built in to the vehicle by the manufacturer.

10.4 CLOSING REMARKS

The very essence of Human Factors research stems from the way in which humans interact with technology in their environment. The information technology revolution is changing the way humans interact within their environments and the information that they have access to. Within the vehicle, this has led to increasing technological advancement. Whilst these technologies may enhance the driver's experience and even provide options to monitor and improve the driving performance, caution is expressed about the additional driver goals that may inhibit drivers from achieving the main driving goal of arriving at their destination safely. The research presented in this book has strived to determine how the wider sociotechnical system interacts to influence the drivers' engagement with technologies in the vehicle that may provide sources of distraction. This has revealed serious concerns for current government policy that states drivers should not be distracted, but does not provide them the necessary support to prevent them from engaging with distractive (nomadic) devices. As the driving task is becoming increasingly 'technology-centric' (Salmon et al., 2012), the prioritisation of creating a safe road environment needs to be the joint focus of the whole sociotechnical system.

This work should aid the application of sociotechnical systems theory to the methods used to assess driver distraction as well as developing the countermeasures. It is hoped that this book shows the relevance of the wider sociotechnical system to the issue of driver distraction and that it will motivate further research to apply methods that reveal the interacting elements of the system and measures that focus not just on the driver but account for the complexly interacting factors that are evident within the system as a whole.

Appendix A
Thematic Framework

Thematic framework of the causal factors relating to the factors that influence the drivers' likelihood of engaging with technological tasks.

Semantic Subthemes	Descriptive Subthemes	Quoted Examples
Driver		
View of self/ behaviour	Anxious	*I find it very difficult to not drive without music because I get quite anxious when I'm on the road, so I use it as a calming mechanism.*
	Ashamed	*I have done that on a motorway I'm ashamed to say.*
	Bad at multitasking	*I just wouldn't be able to multi-task in that way.*
	Didn't think	*Because I didn't really think it through, I just did it.*
	Guilty	*Changing a song/audio track, that's something I'm very guilty of a lot.*
	Lack of technological skill	*Yes, maybe it is my lack of technology skills and my fat fingers.*
	Naughty	*If I am stopped I generally am a little bit more naughty.*
	Poor sense of direction	*I've got a poor sense of direction, so I use sat-navs an awful lot.*
	Scary	*Answer a phone call. Yeah. I do which is actually slightly scary*
	Sneaky	*…then I probably would take a sneaky look at it*
	Stupid	*Just stupid thinking back to it now but we used to, on the way to and back from work*
	Terrible driver	*I am a terrible driver.*
	Too clumsy	*So like, I always mistype stuff or like I hold my phone sometimes and because it's too big I drop it.*
	Too old	*No. I am just one of the – I am probably a different generation, I am too old!*
	Unwise	*It's unwise really*
Driver attitude		
Negative	Distracting	*Because it's – answering the phone is somewhat distracting*
	Got to be crazy	*Yes, I think you'd be crazy to try and enter (laughs) something in to a sat nav*
	Not a good thing to do	*Because, I just think it is the worst thing in the world, I just wouldn't do it, it's terrible.*
	Never	*No, never, never, never, never, never, never, never. Read a text? Honestly I would never do this stuff!*
	Selfish	*…just making a negative impact on someone elses' life while I am just selfishly doing things, so I'm just scared of harming people.*

(Continued)

Appendix A

Semantic Subthemes	Descriptive Subthemes	Quoted Examples
	Stupid	*Yeah no, that sounds stupid.*
	Too Dangerous	*I think I would find it too dangerous.*
	You should be concentrating on the road	*...because you are supposed to be concentrating on your driving not playing*
Positive	Happily do it	*Yes, I would be happy to do that on a motorway and A road.*
	No problem	*I don't see any problem with it personally whatsoever.*
	Should be doing while driving	*It's going to be very similar because it's what I believe I should be doing when I'm driving.*
	Wouldn't mind too much	*So, yeah, probably two, and if I stopped at a light I definitely wouldn't mind looking.*
	Confident	*I feel quite confident in it, the radio is just here, and I just have to put my hand over and just switch it on*
Unnecessary	Don't know why	*I don't know why I do it when I am stopped, because I know that that is illegal*
	Don't want to get involved	*Because I don't want to get involved in all that palaver!*
	It can wait	*That's something – it's just something that can wait until when you get home I think*
	Like being cut-off	*For me, I quite like being cut off when I drive so yeah I am not the type of person that needs to do that.*
	Not worth it	*Yes, too many pedestrians, cyclists, cats, dogs, children, I just think it is not worth it.*
	Phone is for me	*Yes, because the phone is for me, not them. It is having just a bit of attitude!*
	Pointless	*I don't go in for social media on my phone at all whilst driving because it's just pointless.*
	Why would you	*This would be a "no" to all of it because again, I just don't see why you would need to.*
Influence of others	If something happened	*the shame if you did something bad, that everyone would think you are so stupid.*
	Negative of others	*I've seen people taking movies of themselves driving and I think, "How can you be doing that? Because you're looking at your phone, you're not looking at the road at all, what are you doing?"*
	No one else suffers	*I'll wait until I feel kind of like I'm in a comfortable... like there's not too much going on around me so that if anything does happen, the only one that's going to suffer is me, not anyone else.*
	Joining in	*Yeah, yeah. I sort of joined a craze for a little bit.*
	Others seeing you	*Some of it is the public shame, if you did something bad everyone would think you are so stupid.*

(Continued)

Appendix A **209**

Semantic Subthemes	Descriptive Subthemes	Quoted Examples
	Passenger interacts with technology	*I look at it obviously to drive, but I have noticed that when I have people in the car they look at it because they can also figure out where they are, from their world view.*
	Social peer pressure	*The social peer pressure, some people don't think it's a very good thing to do and have told me.*
	Stories of accidents	*My friend's mother got killed by someone being distracted, so this is really important to me.*
Tendency	Can't say I wouldn't	*I can't say I've never done it in my life*
	Do all the time	*So, I do that all the time. I'd say pretty much.*
	Done before	*I have been known to do that*
	Don't do it anymore	*No. I used to check my emails a few years ago while I was working for another company*
	Don't like doing it	*Yes, so on a motorway I don't like doing it but I would do it*
	Don't tend to	*Okay, I have used my phone as a sat nav in the past; I tend not to.*
	Hate to say it	*Well I hate to say it, but I have done this.*
	Hope they wouldn't	*I would like to think I wouldn't take a selfie then that would be where I would be most likely to.*
	I've pulled over	*Nowhere, I just wouldn't do it; I'd pull over and then do it pulled over.*
	Wouldn't think twice	*I wouldn't think twice about verbal communication on either of those.*
Infrastructure		
Perceptions of surrounding environment	Busyness	*Because a major road, there's - I just feel like there's a lot more going on, on a major road.*
	Perceived risk	*I am more likely to do it if it is on a motorway where I would say the risks are somewhat less*
	Familiarity	*…apart from I have to say navigating a junction, especially in an unfamiliar town or parking*
	Required attention	*And maybe less there, very very unlikely for these roads and junctions, it would require a lot more concentration.*
	Other road users	*I think so, maybe when it wouldn't would be residential and urban roads where you have got to navigate traffic calming, you know whatever it might be, someone crossing.*
Road layout	Corners	*Rural roads make me nervous because you just don't know what is going to be around the corner, they make me nervous.*
	Opportunity to stop	*On A roads, urban roads, rural roads and residential roads you can't so much and as I said, the opportunity to stop and actually find something is more convenient*

(Continued)

Appendix A

Semantic Subthemes	Descriptive Subthemes	Quoted Examples
	Straight line	*The motorway is straighter so you can monitor your environment and predict – well maybe predict it a bit better.*
	Turnings	*On an urban road I'd probably say that I'd monitor a route more closely because you're navigating by road names and small turnings which can come up quicker*
	Complicated	*Urban roads...because these are the complicated parts of the journey often, aren't they?*
	Road environment	*Sometime you can along these roads there aren't always great road markings and then, in the picture there you have got high trees and that kind of things,*
	Traffic moving in the same direction	*Yeah, I would do that because to me a motorway, once you are on it, it is all moving in the same direction generally.*
	Consistent	*...whereas when you are on a motorway, you are kind of on a route that is easier to follow. It is more consistent, the route.*
	Road information	*...and the information you're given on the road is actually relatively – it's in advance*
	Size	*Rural road probably, it depends how rural, if it a really really tiny road I probably don't answer it because it is quite nice to be able to hear the road.*
	Lanes	*I find music really distracting if I am parking or need to know which lane I need to be in.*
	Poor visibility	*Yes, they've got very poor visibility*
Road-related behaviour	Speed	*Yes, because we're driving so quickly down those roads*
	Performing a manoeuvre	*I wouldn't do it when I'm manoeuvring through a junction, when I'm driving through a junction*
	Break from looking at the road	*Just sick of staring at the road and so I would just have a quick look whilst I'm stopped at a traffic light.*
	Handbrake on	*Junction it would probably be higher, again because of the whole handbrake on.*
	Not driving	*I would wait until I'm stopped at a junction, instead of doing it when I'm actually driving.*
	Less driving involved	*I would say there is a higher chance on a motorway that I would read a text, because I would feel there was less driving involved*
	Lane changing	*...because you do need to be very aware on a motorway about lane control and changing*
	Long journey	*Okay, if you're on a motorway then yes I would because again it's a long bit of road,*
	Difficult driving task	*Rural again, it's a more difficult driving task.*
	Experience	*I grew up in these rural roads and you really need to be on it, you can turn a corner and there is a horse, you really need to be on it.*

(Continued)

Appendix A   211

Semantic Subthemes	Descriptive Subthemes	Quoted Examples
	Nervous	*...you see I don't like rural road driving, it makes me nervous!*
	Commute to work	*I think because I am generally going slow, I am thinking that this is on my way to work which is generally everything is stopped and you only move about 10ft and then it stops again.*
	More decision	*...an urban road I think is more busy as well so I think the more sort of decisions you've got to make*
Illegality	Know I shouldn't	*I don't know why I do it when I am stopped, because I know that that is illegal still as well but something about the engine running. But I do, if I am being honest.*
Task-road relationship	Slow down	*But on a rural road I'd maybe slow down to do it, but it'd probably be just as likely.*
	Task quality	*Rural roads, I suppose you've got the reception issue, you're less likely to have good reception.*
	Utility	*I think I would need to use it more then, and even maybe in urban roads, because these are the complicated parts of the journey often aren't they?*
	Wait for right conditions	*I think I am likely to be very careful. I can see plenty way ahead, there is nothing happening before I engage.*
	Limited completion	*Yeah it would be stilted, I would probably make the person on the phone aware, say hang on a minute but I would probably sound not as engaged in the conversation.*
	Perceived ability	*Junction I wouldn't because I think you just can't.*
	Not focused on task	*Because I think you'll be watching out and paying more attention to other things*
	Prepare in advance	*Driving through a junction, actually – no I would have checked it before I went through.*
	Consequences	*No, I probably wouldn't because I wouldn't want to have to have the phone call afterwards.*
	Give it a go	*A junction, yes stopped, I'd give it a go.*
	Good opportunity	*Maybe junction, if I stop at a junction I'd be a bit more likely because while I'm stationary I might take the opportunity to just crank it up. I might do that.*
	Time stopped	*Unless I was stopped for a period of time, if you're stopped,*
	Check nothing around you	*You'd just want to check to make sure that there wasn't anything in front of you immediately or around you and then that buys you quite a lot of time on the motorway.*
	Ease of task	*So, on a motorway, yes. I'd enter a number quite easily.*
	Just cruising	*Climate controls, I definitely would on the motorway, as long as you are just cruising along.*
	More time	*Changing a radio station and a media system, motorway very likely for the reasons stated before; you have a lot of time to just flick between them*

(Continued)

Semantic Subthemes	Descriptive Subthemes	Quoted Examples
	Pull over	*Residential I'd just definitely pull over.*
	Task length	*Residential roads, less so. If it was more than just a page that showed up at the start.*
	Use alternative	*Residential … I probably would use a voice thing I think.*
	Cautious	*Rural roads, more cautious but still highly likely,*
	Do it quickly	*Rural roads, yeah, wouldn't mind reading one quickly.*
	Adjust task	*Urban road slightly less likely but probably just turn it off if I had to.*
Task		
Ability to complete	Adapt for driving	*I would probably filter calls more aggressively when I'm driving than I would normally.*
	Consequences of initiating	*No, I probably wouldn't because I wouldn't want to have to have the phone call afterwards,*
	Familiarity	*So certainly at least in my car, I don't have to look at the climate control system to change – because I've had the car for ages, I know where the switches are.*
	Limited completion of task	*If I was manoeuvring I would recognise the text has come through but I wouldn't read it to the extent that I would have taken it in, so I would glance over.*
	Preset options	*This one is pre-set channels on the radio and pre-set channels and I only have to touch one button.*
	Couldn't do it	*Because I can't drive and enter a destination at the same time because I would probably kill myself.*
	I could do it	*If I can find a number from an address book, I can press a button to answer a call on the motorway.*
Complexity	Cognitive processing required	*Because you are not simply taking your eyes off the road to look at the radio or the heating or something. Its reading something and then processing that.*
	Concentration required	*I think ultimately I am going to be happy to do them on all of them, it doesn't really require much attention being taken away from the driving.*
	Difficulty of task	*Just because it's a bit more fiddly*
	Ease of task	*Just because if you're clicking the air-conditioning button or changing the radio station or whatever, all of those you can do fairly easily*
	Phone unlock	*If you have to unlock the phone screen or whatever, it is not as simple – well it is quite distracting…*
Desirability	Effort	*It's exceptionally lazy; it really is an exercise in laziness. So again, I would be very likely to.*
	No alternative	*On phone. yes, I have had to actually, when my normal system's gone ka-blink, I've had to.*
	Preference for alternative	*So, finding a number is probably nil for me because I would use the voice activated command, so I don't have to look away which is quite good.*

(*Continued*)

Appendix A

Semantic Subthemes	Descriptive Subthemes	Quoted Examples
	Reliability	*I would make a call usually only with the headset usually because it is more reliable.*
	Technology quality	*Yes, it would probably be quite high. I don't like these systems because they never work.*
	Trust	*That would come down to trust. If I could trust it would recognise my voice I would do it all the time.*
	Use outside car	*I don't really use my phone very much anyway so it's never been something that I have felt I needed.*
	Utility	*I've got a poor sense of direction, so I use sat navs an awful lot.*
Duration	Length of text	*If it's just a couple of words then you can read it quite easily; if it's a long text you might not read it.*
	Time	*Generally, I am driving at a constant speed, so I think yeah I do have time...*
Interaction	Button presses	*It's only one button to press, so that's not an issue.*
	Can see task	*Yeah, that's easy, there's one knob and can see it.*
	Don't have to look	*And I would never take my eyes off the road doing that because I just don't need to; I know where the button is.*
	Don't touch it	*Most of the time it's just on shuffle though. So, I don't really need to touch it.*
	Eyes off road	*No, I'd probably – because you have to take your eyes off the road.*
	Eyes still on road	*I can see it and I can do it while keeping my eye on the road, so I would say that is something I might do.*
	Hands on the wheel	*Because it is on the steering wheel I don't see it as a major distraction, compared to if I didn't have that and it was me taking a hand off the wheel to change station it would probably be kind of different.*
	Human-machine interface	*They went through a stage where they had up/down buttons for volume...on the console in the middle and that was really annoying and they've gone back to knobs now and if it's a knob, I would do it whenever.*
	Integrating with the driving task	*I think it is quite easy, again like using a phone I would do it staggered touch something look up, touch something else, look up.*
	Location of device	*It's usually when it's in the little dashboard holder because it's right there. I think – so often usually I don't put my phone – I'll only put my phone in that thing if I need to charge it; otherwise I just put it in the thing by the handbrake and if it's in there I don't read it*
	Mind off the road	*If I start talking I might just start to imagine my eyes, they literally don't see anything in front of me but just what I'm thinking.*

(*Continued*)

214   Appendix A

Semantic Subthemes	Descriptive Subthemes	Quoted Examples
	Processes involved	*No because that would involve me having to search through for – going on to the internet browser, finding the website, entering the – and no, no.*
Engagement regulation	Car doesn't allow it	*Navigation? Whilst driving on a motorway! Well my car doesn't let me do it, for one thing.*
	Conscious decision	*I would probably look to see who it was and then I would make a decision if I was going to pull over and phone those people back or just ignore it*
	Curiosity	*if I'm expecting to meet someone, if I was going to do something and it was on the seat or something, I would think, "Oh," your curiosity would be like, "Oh, I wonder who that's from?*
	Do when not driving	*So, I will always figure out what I'm going to listen to and set it going before I leave.*
	Don't use device/task	*…it has, it has got hands free technology, but I don't use it.*
	Drawn in	*Yes it is not audio because I find that too distracting, I get too involved if it is a story.*
	Ignore	*If anything were to happen on the phone, for example I hadn't got it connected to my Bluetooth and I got a call, I would ignore it, which I've done on a number of occasions*
	Reduce boredom	*I think it's just like driving's quite boring, isn't it, on the whole, so you tend to … a phone's quite an easy thing to distract you or just sort of get distracted, but like, you know, take away from all the boredom really.*
	Response time	*Yes, so across all of them I would think it would all be context driven, whether I accept the call, if I think I could actually press the button to accept the call quickly enough.*
	Triggered response	*"Reading a text", you see I would read a text just because of the nature of the fact that it flashes up on your phone.*
Context		
Journey context	Familiarity	*I am now wondering if I am in a strange city, I would be less likely to mess around because I don't know where I am going.*
	Journey length	*So, it is who, again length of journey, if I am going to stop soon I probably wouldn't, But I would say I was driving.*
	Journey type	*Check your email. Yeah, I do, yeah. On my way home from work.*

(Continued)

Appendix A

Semantic Subthemes	Descriptive Subthemes	Quoted Examples
	Time of day or night	*Even when you're on a motorway, if my phone went off and it's dark, then I don't like to look at it just because then your face is all lit up and someone else would know that you were looking at a phone and holding it.*
Road context	Across all road types	*So that is always, I would always do that, so across all road types.*
	Busyness environment	*I think it would be situational dependent, just how busy is it? I think … it would depend I think.*
	Concentration required from driving	*And also, if I wasn't so busy thinking about the driving that I realised that I couldn't actually speak to the person.*
	Feels safe/ comfortable	*…so again, it would be on whether I felt it was safe or not.*
Task context	Urgency/Importance	*It's stuff when I actually feel like I need to send a message quickly, so if I've agreed to come home at a certain time and I'm running late for instance.*
	Expecting something	*I don't like doing it, but then if you're expecting a call or… and I mean I think if it was a call I'm expecting I would probably answer it*
	Necessity	*If I needed to I would, but I don't do it that often.*
	Required info to hand	*There's a chance I would have a go at putting in the – and I had the postcode to hand, there's a chance I would do it.*
	In a hurry	*If I was on a motorway going straight and I really needed to be somewhere, I was in a hurry, there's a chance*
	Who they are communicating with	*So, answer the call, I am actually quite unlikely to answer actually it would depend on who it was. If it is my boss then that is different, so it depends.*

Appendix B
Systemic Factors Influencing Distraction

TABLE B.1

Summary of Key Factors Increasing and Decreasing Drivers' Engagement with Different Tasks

Task Scenario			
Text	**Call**	**Destination**	**Song**
Factors Increasing Engagement			
Context	*Context*	*Task*	*Task*
• Expecting a text	• Importance of the phone call	• Enter in stages (glance back to road)	• Don't need to look
• The importance of the message	*Task*	• Use voice input	• Know where the buttons are
Task	• Connect to hands-free/speaker	• Use remembered destinations within the system	• Easy to take one hand off the steering wheel
• Phone placed in line of sight, for example, on dashboard	• Voice dial	• Use voice entry	• Very quick and easy to do safely
• Notifications flashes up on screen (reflex reaction)	• Small glances to scroll phone book		• It's an automatic action
• Read first lines to determine urgency/ who its from	• If phone is in line of sight		• Use steering wheel buttons
• Single touch/ Easily accessible	• Thumb print phone unlock		• Use pre-set buttons
• Place phone on lap	• Number saved to contacts		• Single button presses
• Gradual glances away from road	• If single button-press		• Only one quick glance
• Short texts	• Quick and easy to achieve		• Don't have to take eyes off road
• Touch unlock (easier than pin)	• If already connected, would continue to talk (even in high-demand areas where call may not be initiated)		• Don't like bad songs
			• Know where the buttons are
			• Don't have to think about it

(Continued)

TABLE B.1 (*Continued*)

Summary of Key Factors Increasing and Decreasing Drivers' Engagement with Different Tasks

		Task Scenario	
Text	**Call**	**Destination**	**Song**
Factors Reducing Engagement			
Context	*Driver*	*Driver*	*Driver*
• Not urgent enough	• Never use phone while driving	• Don't want to take hands off the wheel	• Not even thinking about it
Driver	• Never going to be important enough	• Need glasses to read the device	*Task*
• Doesn't need immediate action	• Not capable of doing task		• Too much faff to change, easier to turn it off
• Place phone out of sight	• It can wait	*Task*	• If listening through phone – need to unlock phone
• Would not use phone at all in vehicle	*Task*	• Poorly designed feature	• Need hands on the steering wheel
Task	• Lengthy menus to scroll through	• Too long/eyes off road for too long	• If have to go into menus
• Needing to touch the phone	• Unlocking the phone	• Too many key presses	
• Phone placed out of sight	• Manual/full number entry	• Complex interaction (more than 1 press)	
• Need to enter phone pin code	• Scrolling through is too complex		
• Long messages	• Need hands on the wheel		
• If scrolling is involved			
• Not enough time to read it			

TABLE B.2

Summary of Key Factors Increasing and Decreasing Drivers' Engagement across Different Road Types

Task Scenario

Motorway	A Road	Roundabout

Factors Increasing Engagement

Motorway	A Road	Roundabout
Context	*Context*	*Infrastructure*
• Familiarity with the road	• Quiet road	• Stopped, for example, at light, with handbrake on
• On the road for a long time	• Familiarity with the road	• During waiting zone before the roundabout
• Low traffic levels	• Going slowly	• If a queue at the roundabout
Infrastructure	*Infrastructure*	• Go around roundabout again to get caught by a light and enter destination
• Can see far ahead	• Low-demand road	• Straighter sections of the roundabout
• Stable/predictable environment	• Straight road/good visibility	
• Consistent speed	• Slow stop/start traffic	
• Low-risk environment	• Boring road	
• Move to slow lane	• Rural areas/less built up	
• Straight segments of road	• Wide road	
• Limited attention required to road	• Can't stop/pull over	
Other road users	• No junctions	
• Increase distance to lead vehicles	*Other road users*	
• Low levels of traffic around	• Predictable road users	
• Road users keeping their trajectory	• Leave gap to vehicle in front	

(Continued)

TABLE B.2 (*Continued*)

Summary of Key Factors Increasing and Decreasing Drivers' Engagement across Different Road Types

Task Scenario		
Motorway	**A Road**	**Roundabout**
Factors Reducing Engagement		
Infrastructure	*Infrastructure*	*Context*
• Need to check lanes a lot	• Residential/built-up areas	• Don't know where the exit is/not familiar
• Approaching a junction	• Junctions with merging traffic	• Driving demand is too high
• Bends and obscured line of sight	• Opportunity to pull over	*Infrastructure*
• Things happen quickly	• Wait for a red light	• Wait until after the roundabout
• Changing lanes	• Small/narrow road	• Big manoeuvre/too much steering
• Can't pull over on the motorway	• If turning/manoeuvring the vehicle	• Needs a lot of attention, especially trying to pull out.
Other road users	• Windy roads	• Need to change lanes
• Large vehicles around	• Wait until after a bend	• Unpredictable road environment
• Lots of traffic	*Other road users*	• Good judgement needed to enter roundabout
• Fast traffic from behind	• Near cyclists/pedestrians	• Blind spots
	• Lots of oncoming traffic	• Concentration to stay in the right lane
	• Parked cars, watch for pedestrians	*Other road users*
	• Lorry on the road	• Looking for oncoming cars/hazards
	• Erratic driver ahead	• Lots of cars changing lanes
	• Pedestrians may see you using the device (anti-social)	• Other traffic monitoring

Key factors influencing engagement likelihood across road types, as drawn from the drivers discussions, are presented. These are categorised into the systemic themes identified across the interviews and driving studies.

References

Ajzen, I. (1991). The theory of planned behaviour. *Organizational Behaviour and Human Decision Processes*, 50(2), 179–211.

Alliance of Automobile Manufacturers (2006). Statement of Principles, Criteria and Verification Procedures on Driver-Interactions with Advanced In-Vehicle Information and Communication Systems, *Driver Focus-Telematics Working Group*. June, Washington, DC.

Allison, C. K., Parnell, K. J., Brown, J. W., & Stanton, N. A. (2017). Modelling the Real World Using STISIM Drive® Simulation Software: A Study Contrasting High and Low Locality Simulations. In *International Conference on Applied Human Factors and Ergonomics* (906–915). Springer, Cham.

Alm, H., & Nilsson, L. (1994). Changes in driver behaviour as a function of handsfree mobile phones—A simulator study. *Accident Analysis & Prevention*, 26(4), 441–451.

Alm, H., & Nilsson, L. (1995). The effects of a mobile telephone task on driver behaviour in a car following situation. *Accident Analysis & Prevention*, 27(5), 707–715.

Altman, D. G. (1990). *Practical Statistics for Medical Research*. Chapman & Hall, London.

Apple (2017). How to use the Do Not Disturb while driving. https://support.apple.com/en-gb/HT208090. Accessed 02/10/2017.

Atchley, P., Atwood, S., & Boulton, A. (2011). The choice to text and drive in younger drivers: Behaviour may shape attitude. *Accident Analysis & Prevention*, 43(1), 134–142.

Atchley, P., Hadlock, C., & Lane, S. (2012). Stuck in the 70s: The role of social norms in distracted driving. *Accident Analysis & Prevention*, 48, 279–284.

Axon, S., Speake, J., & Crawford, K. (2012). 'At the next junction, turn left': Attitudes towards Sat Nav use. *Area*, 44(2), 170–177.

Baber, C., Stanton, N. A., Atkinson, J., McMaster, R., & Houghton, R. J. (2013). Using social network analysis and agent-based modelling to explore information flow using common operational pictures for maritime search and rescue operations. *Ergonomics*, 56(6), 889–905.

Banks, V. A., & Stanton, N. A. (2015). Contrasting models of driver behaviour in emergencies using retrospective verbalisations and network analysis. *Ergonomics*, 58(8), 1337–1346.

Banks, V. A., & Stanton, N. A. (2016). Keep the driver in control: Automating automobiles of the future. *Applied Ergonomics*, 53, 389–395.

Banks, V. A., Stanton, N. A., Burnett, G., & Hermawati, S. (2018). Distributed cognition on the road: Using EAST to explore future road transportation systems. *Applied Ergonomics*, 68, 258–266.

Banks, V. A., Stanton, N. A., & Harvey, C. (2014a). Sub-systems on the road to vehicle automation: Hands and feet free but not 'mind' free driving. *Safety Science*, 62, 505–514.

Banks, V. A., Stanton, N. A., & Harvey, C. (2014b). What the drivers do and do not tell you: Using verbal protocol analysis to investigate driver behaviour in emergency situations. *Ergonomics*, 57(3), 332–342.

Bayliss, D. (2009). Accident Trends by Road type. Motoring Towards 2050—Roads and Reality Background Paper No. 9. Royal Automobile Club Foundation. www.racfoundation.org/assets/rac_foundation/content/downloadables/roads%20and%20reality%20-%20bayliss%20-%20accident%20trends%20by%20road%20type%20-%20160309%20-%20background%20paper%209.pdf. Accessed 01/08/2017.

BBC News (2013). Cyclist Anthony Hilson death: Victoria McClure jailed. www.bbc.co.uk/news/uk-england-berkshire-23904694. Accessed 21/08/2015.

BBC (2017). Thousands of Drivers Caught Despite Mobile Crackdown. www.bbc.co.uk/news/business-40079382. Accessed 31/05/2017.

Beanland, V., Fitzharris, M., Young, K. L., & Lenné, M. G. (2013). Driver inattention and driver distraction in serious casualty crashes: Data from the Australian National Crash In-depth Study. *Accident Analysis & Prevention*, 54, 99–107.

Bella, F. (2008). Driving simulator for speed research on two-lane rural roads. *Accident Analysis & Prevention*, 40(3), 1078–1087.

Bittner, A., Simsek Jr., O., Levison, W., & Campbell, J. (2002). On-road versus simulator data in driver model development driver performance model experience. *Transportation Research Record: Journal of the Transportation Research Board*, 1803, 38–44.

Blomquist, G. (1986). A utility maximization model of driver traffic safety behaviour. *Accident Analysis & Prevention*, 18(5), 371–375.

Boer, E. R., & Hoedemaeker, M. (1998). Modeling driver behavior with different degrees of automation: A hierarchical decision framework of interacting mental models. In *Conference on Human Decision Making and Manual Control*. LAMIH, Valenciennes.

Bojko, A. A. (2009). Informative or misleading? Heatmaps deconstructed. In *International Conference on Human-Computer Interaction*, 30–39. Springer, Berlin.

Boyatzis, R. E. (1998). *Transforming Qualitative Information: Thematic Analysis and Code Development*. Sage, London.

Branford, K., Hopkins, A., & Naikar, N. (2009). Guidelines for AcciMap analysis. In Hopkins, A. (Ed.) *Learning from High Reliability Organisations*. CCH Australia Ltd, Sydney, NSW, 193–212.

Braun, V., & Clarke, V. (2006). Using thematic analysis in psychology. *Qualitative Research in Psychology*, 3(2), 77–101.

Brodsky, W. (2002). The effects of music tempo on simulated driving performance and vehicular control. *Transportation Research Part F: Traffic Psychology and Behaviour*, 4(4), 219–241.

Brodsky, W., & Slor, Z. (2013). Background music as a risk factor for distraction among young-novice drivers. *Accident Analysis & Prevention*, 59, 382–393.

Brookhuis, K. A., de Vries, G., & de Waard, D. (1991). The effects of mobile telephoning on driving performance. *Accident Analysis & Prevention*, 23(4), 309–316.

Brookhuis, K. A., De Waard, D., & Janssen, W. H. (2001). Behavioural impacts of advanced driver assistance systems—An overview. *European Journal of Transport and Infrastructure Research*, 1(3), 245–253.

Brown, I. D. (1965). Effect of a car radio on driving in traffic. *Ergonomics*, 8(4), 475–479.

Brown, I. D., Tickner, A. H., & Simmonds, D. C. (1969). Interference between concurrent tasks of driving and telephoning. *Journal of Applied Psychology*, 53(5), 419.

Bunn, T. L., Slavova, S., Struttmann, T. W., & Browning, S. R. (2005). Sleepiness/fatigue and distraction/inattention as factors for fatal versus nonfatal commercial motor vehicle driver injuries. *Accident Analysis & Prevention*, 37(5), 862–869.

Burnett, G. E. (2000). Usable vehicle navigation systems: Are we there yet. *Vehicle Electronic Systems-European Conference and Exhibition*, ERA Technology Ltd, 3.1.1–3.1.11.

Burnett, G. (2009). In Zaphiris, P., & Ang, C.S. (Eds.) *Human Computer Designing and Evaluating In-car User Interfaces Interaction: Concepts, Methodologies, Tools, and Applications*. IGI Global, New York, 217–235.

Burnett, G. E., Irune, A., & Mowforth, A. (2007). Driving simulator sickness and validity: How important is it to use real car cabins? *Advances in Transportation Studies*, 13, 33–42.

Burnett, G. E., & Joyner, S. M. (1997). An assessment of moving map and symbol-based route guidance systems. In Ian Noy, Y. (Ed.) *Ergonomics and Safety of Intelligent Driver Interfaces*. Lawrence Erlbaum Associates, Mahwah, NJ, 115–136.

References

223

Burnett, G. E., Summerskill, S. J., & Porter, J. M. (2004). On-the-move destination entry for vehicle navigation systems: Unsafe by any means? *Behaviour & Information Technology*, 23(4), 265–272.

Burns, P. C., Parkes, A., Burton, S., Smith, R. K., & Burch, D. (2002). How dangerous is driving with a mobile phone? Benchmarking the impairment to alcohol. *TRL report TRL547*. TRL Limited, Berkshire.

Burstein, P. (2003). The impact of public opinion on public policy: A review and an agenda. *Political Research Quarterly*, 56(1), 29–40.

Byrne, E. A., & Parasuraman, R. (1996). Psychophysiology and adaptive automation. *Biological Psychology*, 42(3), 249–268.

Cacciabue, P. C. (2013). *Guide to Applying Human Factors Methods: Human Error and Accident Management in Safety-Critical Systems*. Springer Science & Business Media, London.

Caird, J. K., Johnston, K. A., Willness, C. R., & Asbridge, M. (2014). The use of meta-analysis or research synthesis to combine driving simulation or naturalistic study results on driver distraction. *Journal of Safety Research*, 49, 91.e1–96.e1.

Campbell, J. L., Quincy, C., Osserman, J., & Pedersen, O. K. (2013). Coding in-depth semi-structured interviews: Problems of unitization and inter-coder reliability and agreement. *Sociological Methods & Research*, 42(3), 294–320.

Campbell, S. W. (2006). Perceptions of mobile phones in college classrooms: Ringing, cheating, and classroom policies. *Communication Education*, 55(3), 280–294.

Cantin, V., Lavallière, M., Simoneau, M., & Teasdale, N. (2009). Mental workload when driving in a simulator: Effects of age and driving complexity. *Accident Analysis & Prevention*, 41(4), 763–771.

Carsten, O., Kircher, K., & Jamson, S. (2013). Vehicle-based studies of driving in the real world: The hard truth? *Accident Analysis & Prevention*, 58, 162–174.

Carsten, O., & Merat, N. (2015, November). Protective or Not? In *4th International Conference on Driver Distraction & Inattention*, Sydney, NSW.

Cassano-Piche, A. L., Vicente, K. J., & Jamieson, G. A. (2009). A test of Rasmussen's risk management framework in the food safety domain: BSE in the UK. *Theoretical Issues in Ergonomics Science*, 10(4), 283–304.

Cavaye, A. L. (1996). Case study research: A multi-faceted research approach for IS. *Information Systems Journal*, 6(3), 227–242.

Charlton, S. G., & Starkey, N. J. (2011). Driving without awareness: The effects of practice and automaticity on attention and driving. *Transportation Research Part F: Traffic Psychology and Behaviour*, 14(6), 456–471.

Chen, H. Y. W., & Donmez, B. (2016). What drives technology-based distractions? A structural equation model on social-psychological factors of technology-based driver distraction engagement. *Accident Analysis & Prevention*, 91, 166–174.

Chen, H. Y. W., Donmez, B., Hoekstra-Atwood, L., & Marulanda, S. (2016). Self-reported engagement in driver distraction: An application of the Theory of Planned Behaviour. *Transportation Research Part F: Traffic Psychology and Behaviour*, 38, 151–163.

Cnossen, F., Meijman, T., & Rothengatter, T. (2004). Adaptive strategy changes as a function of task demands: A study of car drivers. *Ergonomics,* 47(2), 218–236.

Cnossen, F., Rothengatter, T., & Meijman, T. (2000). Strategic changes in task performance in simulated car driving as an adaptive response to task demands. *Transportation Research Part F: Traffic Psychology and Behaviour*, 3(3), 123–140.

Cohen, D., & Crabtree, B. (2006) Qualitative Research Guidelines Project. www.qualres.org/HomeSemi-3629.html. Accessed 22/04/2017.

Commission of the European Communities (2006). Commission Recommendation of 22 December 2006 on safe and efficient in-vehicle information and communication systems: Update of the European Statement of Principles on human machine interface. *Official Update of the European Union*. 78. EC.

Commission of the European Communities (2008). Commission Recommendation of 26 May 2008 on Safe and Efficient In-Vehicle Information and Communication Systems; Update of the European Statement of Principles on Human-Machine Interface.

Cooper, P. J., & Zheng, Y. (2002). Turning gap acceptance decision-making: The impact of driver distraction. *Journal of Safety Research*, 33(3), 321–335.

Cummings, P., Koepsell, T. D., Moffat, J. M., & Rivara, F. P. (2001). Drowsiness, counter-measures to drowsiness, and the risk of a motor vehicle crash. *Injury Prevention*, 7(3), 194–199.

Darke, P., Shanks, G., & Broadbent, M. (1998). Successfully completing case study research: Combining rigour, relevance and pragmatism. *Information Systems Journal*, 8(4), 273–289.

Davidsson, S., & Alm, H. (2014). Context adaptable driver information–Or, what do whom need and want when? *Applied Ergonomics*, 45(4), 994–1002.

de Waard, D., Schepers, P., Ormel, W., & Brookhuis, K. (2010). Mobile phone use while cycling: Incidence and effects on behaviour and safety. *Ergonomics*, 53(1), 30–42.

Dekker, S. W. (2002). Reconstructing human contributions to accidents: The new view on error and performance. *Journal of Safety Research*, 33(3), 371–385.

Dekker, S. W. (2011). *Drift into Failure: From Hunting Broken Components to Understanding Complex Systems*. Ashgate Publishing Ltd., Aldershot, UK.

Dekker, S. W. (2014). *The Field Guide to Understanding 'Human Error'*. Ashgate Publishing, Ltd., Aldershot, UK.

Denzin, N. K., & Lincoln, Y. S. (2011). *The Sage Handbook of Qualitative Research*. Sage, Thousand Oaks, CA.

Department for Transport (2000). Tomorrow's Roads: Safer for Everyone. Online resource: http://webarchive.nationalarchives.gov.uk/20100202151921/http://www.Department forTransport.gov.uk/pgr/roadsafety/strategytargetsperformance/tomorrowsroadssafer-foreveryone. Accessed 02/19/2017.

Department for Transport (2015a). Reported Road Casualties Great Britain: 2015. Annual Report. www.gov.uk/government/uploads/system/uploads/attachment_data/file/568484/rrcgb-2015.pdf. Accessed 28/04/2017.

Department for Transport (2015b). 2010 to 2015 government policy: Road network and traffic. May. The Stationary Office, London.

Department for Transport (2015c). The Pathway to Driverless Cars: A Detailed Review of Regulations for Automated Vehicle Technologies. www.gov.uk/government/uploads/system/uploads/attachment_data/file/401565/pathway-driverless-cars-main.pdf. Accessed 01/02/2018.

Department for Transport (2016a). Consultation on changes to the fixed penalty notice and penalty points for the use of a hand-held mobile phone whilst driving. The Stationary Office, London.

Department for Transport (2016b). Reported Road casualties Great Britain: 2015 Annual Report. September. London. www.gov.uk/government/uploads/system/uploads/attach-ment_data/file/568484/rrcgb-2015.pdf. Accessed 31/05/2017.

Department for Transport (2017). Increasing Mobile Phone FFPN and Penalty Points for the Offence of using a Mobile Phone Whilst Driving. Hand-Held Mobile Phones: Changes to Penalties for Use Whilst Driving. www.gov.uk/government/uploads/system/uploads/attachment_data/file/565100/mobile-phones-driving-consultation-impact-assessment.pdf. Accessed 31/05/2017.

Dingus, T. A., Guo, F., Lee, S., Antin, J. F., Perez, M., Buchanan-King, M., & Hankey, J. (2016). Driver crash risk factors and prevalence evaluation using naturalistic driving data. *Proceedings of the National Academy of Sciences*, 113(10), 2636–2641.

Dingus, T. A., Klauer, S., Neale, V., Petersen, A., Lee, S., Sudweeks, J., & Gupta, S. (2006). The 100-car naturalistic driving study, Phase II-results of the 100-car field experiment. DOT HS 810 593. April.

References

Dixon, S. M., Nordvall, A. C., Cukier, W., & Neumann, W. P. (2017). Young consumers' considerations of healthy working conditions in purchasing decisions: A qualitative examination. *Ergonomics*, 60(5), 601–612.

Donmez, B., Boyle, L. N., & Lee, J. D. (2008). Mitigating driver distraction with retrospective and concurrent feedback. *Accident Analysis & Prevention*, 40(2), 776–786.

Dogan, E., Steg, L., & Delhomme, P. (2011). The influence of multiple goals on driving behavior: The case of safety, time saving, and fuel saving. *Accident Analysis & Prevention*, 43(5), 1635–1643.

Dong, Y., Hu, Z., Uchimura, K., & Murayama, N. (2011). Driver inattention monitoring system for intelligent vehicles: A review. *IEEE Transactions on Intelligent Transportation Systems*, 12(2), 596–614.

Dorf, R. C. (2001). *Technology, Humans, and Society: Toward a Sustainable World*. Academic Press, San Diego, CA.

Drews, F. A., & Strayer, D. L. (2008). 11 cellular phones and driver distraction. In Regan, M. A., Lee, J. D., & Young, K. (Eds.) *Driver Distraction: Theory, Effects, and Mitigation*. CRC Press, Boca Raton, FL, 169–190.

Driver and Vehicle Licensing Agency (2015). FOI release: How many people hold driving licenses in the UK. www.gov.uk/government/uploads/system/uploads/attachment_data/file/397430/FOIR4341_How_many_people_hold_licences_in_the_UK.pdf. Accessed 27/06/2017.

Drury, C., Abussaud, Z., Allison, G., Bhindi, E., Bustard, J., Chamberlain, J., Fujino, M., Green, H., Harding, L., Ironside, R., Judge, L., Kerse, K., Laing, E., Liu, B., & Judge, L. (2012). Mobile Phone Use While Driving After a New Law: Observational Study. www.otago.ac.nz/wellington/otago041209.pdf. Accessed 02/10/2017.

Dukic, T., Ahlstrom, C., Patten, C., Kettwich, C., & Kircher, K. (2013). Effects of electronic billboards on driver distraction. *Traffic Injury Prevention*, 14(5), 469–476.

Dulisse, B. (1997). Methodological issues in testing the hypothesis of risk compensation. *Accident Analysis & Prevention*, 29(3), 285–292.

Eason, K. D. (1988). *Information Technology and Organisational Change*. CRC Press, Boca Raton, FL.

Edwards, W. (1954). The theory of decision making. *Psychological Bulletin*, 51(4), 380.

Elvik, R. (1999). Can injury prevention efforts go too far? Reflections on some possible implications of Vision Zero for road accident fatalities. *Accident Analysis & Prevention*, 31(3), 265–286.

Engelberg, J. K., Hill, L. L., Rybar, J., & Styer, T. (2015). Distracted driving behaviors related to cell phone use among middle-aged adults. *Journal of Transport & Health*, 2(3), 434–440.

Eost, C., & Galer Flyte, M. (1998). An investigation into the use of the car as a mobile office. *Applied Ergonomics*, 29, 383–388.

Ercikan, K., & Roth, W. M. (2006). What good is polarizing research into qualitative and quantitative? *Educational Researcher*, 35(5), 14–23.

Eriksson, A., Banks, V. A., & Stanton, N. A. (2017). Transition to manual: Comparing simulator with on-road control transitions. *Accident Analysis & Prevention*, 102, 227–234.

Ericsson, K. A., & Simon, H. A. (1993). *Protocol Analysis*. MIT Press, Cambridge.

EURO NCAP (2017). 2025 Roadmap. In pursuit of vision zero. https://cdn.euroncap.com/media/30700/euroncap-roadmap-2025-v4.pdf. Accessed 05/01/2017.

Fagnant, D. J., & Kockelman, K. (2015). Preparing a nation for autonomous vehicles: Opportunities, barriers and policy recommendations. *Transportation Research Part A: Policy and Practice*, 77, 167–181.

Fraedrich, E., Beiker, S., & Lenz, B. (2015). Transition pathways to fully automated driving and its implications for the sociotechnical system of automobility. *European Journal of Futures Research*, 3(1), 11.

Fuller, R. (2000). The task-capability interface model of the driving process. *Recherche-Transports-Sécurité*, 66, 47–57.

Furui, S. (2010). History and development of speech recognition. In Chen, F., & Jokinen, K. (Eds.) *Speech Technology: Theory and Applications*. Springer Science & Business Media, New York, 1–18.

Funkhouser, D., & Sayer, J. (2012). Naturalistic census of cell phone use. *Transportation Research Record: Journal of the Transportation Research Board*, 2321, 1–6.

Gardner, B., & Abraham, C. (2007). What drives car use? A grounded theory analysis of commuters' reasons for driving. *Transportation Research Part F: Traffic Psychology and Behaviour*, 10(3), 187–200.

Gardner, M., & Steinberg, L. (2005). Peer influence on risk taking, risk preference, and risky decision making in adolescence and adulthood: An experimental study. *Developmental Psychology*, 41(4), 625.

Gillham, B. (2000). *Case Study Research Methods*. Bloomsbury Publishing, London, UK.

Glaser, B., & Strauss, A. (1967). *The Discovery Grounded Theory: Strategies for Qualitative Inquiry*. Aldine, Chicago, IL.

Glaser, B. G. (2001). *The Grounded Theory Perspective: Conceptualization Contrasted with Description*. Sociology Press, Mill Valley, CA.

Godley, S. T., Triggs, T. J., & Fildes, B. N. (2002). Driving simulator validation for speed research. *Accident Analysis & Prevention*, 34(5), 589–600.

Gonçalves, J., & Bengler, K. (2015). Driver state monitoring systems–Transferable knowledge manual driving to HAD. *Procedia Manufacturing*, 3, 3011–3016.

Goodman, M. J., Tijerina, L., Bents, F. D., & Wierwille, W. W. (1999). Using cellular telephones in vehicles: Safe or unsafe? *Transportation Human Factors*, 1(1), 3–42.

Gordon, C. P. (2008). Crash studies of driver distraction. In Regan, M. A., Lee, J. D., & Young, K. L. (Eds.) *Driver Distraction: Theory, Effects and Mitigation*, CRC Press, Boca Raton, FL, 281–304.

Green, P. (1999). Estimating compliance with the 15-second rule for driver-interface usability and safety. In *Proceedings of the Human Factors and Ergonomics Society Annual Meeting*, 43(18), 987–991, Sage, Los Angeles, CA.

Green, P. (2004). Driver Distraction, Telematics Design, and Workload Managers: Safety Issues and Solutions (No. 2004-21-0022). SAE Technical Paper.

Green, P., Levison, W., Paelke, G., & Serafin, C. (1995). Preliminary human factors design guidelines for driver information systems (Tech. Rep. No. FHWA-RD-94-087). U.S. Government Printing Office, Washington, DC.

Groeger, J. A. (2000). *Understanding Driving: Applying Cognitive Psychology to a Complex Everyday Task*. Psychology Press, Hove, UK.

Haigney, D. E., Taylor, R. G., & Westerman, S. J. (2000). Concurrent mobile (cellular) phone use and driving performance: Task demand characteristics and compensatory processes. *Transportation Research Part F: Traffic Psychology and Behaviour*, 3(3), 113–121.

Hancock, P. A., Hancock, G. M., & Warm, J. S. (2009). Individuation: The N = 1 revolution. *Theoretical Issues in Ergonomics Science*, 10(5), 481–488.

Hancock, P. A., Lesch, M., & Simmons, L. (2003). The distraction effects of phone use during a crucial driving maneuver. *Accident Analysis & Prevention*, 35(4), 501–514.

Hancock, P. A., Mouloua, M., & Senders, J. W. (2009). On the philosophical foundations of the distracted driver and driving distraction. In Regan, M. A., Lee, J. D., & Young, K. L. (Eds.) *Driver Distraction. Theory, Effects and Mitigation*. CRC Press, Boca Raton, FL, 11–30.

Hancox, G., Richardson, J., & Morris, A. (2013). Drivers' willingness to engage with their mobile phone: The influence of phone function and road demand. *IET Intelligent Transport Systems*, 7(2), 215–222.

References

Hanowski, R. J., Perez, M. A., & Dingus, T. A. (2005). Driver distraction in long-haul truck drivers. *Transportation Research Part F: Traffic Psychology and Behaviour*, 8(6), 441–458.

Harbluk, J. L., Burns, P. C., Lochner, M., & Trbovich, P. L. (2007). Using the lane-change test (LCT) to assess distraction: Tests of visual-manual and speech-based operation of navigation system interfaces. In *Proceedings of the 4th International Driving Symposium on Human Factors in Driver Assessment, Training, and Vehicle Design*, 9, 16–22, Stevenson, WA.

Harbluk, J. L., Noy, Y. I., Trbovich, P. L., & Eizenman, M. (2007). An on-road assessment of cognitive distraction: Impacts on drivers' visual behavior and braking performance. *Accident Analysis & Prevention*, 39(2), 372–379.

Harms, L., & Patten, C. (2003). Peripheral detection as a measure of driver distraction. A study of memory-based versus system-based navigation in a built-up area. *Transportation Research Part F: Traffic Psychology and Behaviour*, 6(1), 23–36.

Hart, S. G., & Staveland, L. E. (1988). Development of NASA-TLX (Task Load Index): Results of empirical and theoretical research. *Advances in Psychology*, 52, 139–183.

Harvey, C., Stanton, N. A., Pickering, C. A., McDonald, M., & Zheng, P. (2011a). Context of use as a factor in determining the usability of in-vehicle devices. *Theoretical Issues in Ergonomics Science*, 12(4), 318–338.

Harvey, C., Stanton, N. A., Pickering, C. A., McDonald, M., & Zheng, P. (2011b). A usability evaluation toolkit for in-vehicle information systems (IVISs). *Applied Ergonomics*, 42(4), 563–574.

Harvey, C., Stanton, N. A., Pickering, C. A., McDonald, M., & Zheng, P. (2011c). In-vehicle information systems to meet the needs of drivers. *International Journal of Human–Computer Interaction*, 27(6), 505–522.

Harvey, C., & Stanton, N. A. (2013). *Usability Evaluation for In-Vehicle Systems*. CRC Press, Boca Raton, FL.

He, J., Becic, E., Lee, Y. C., & McCarley, J. S. (2011). Mind wandering behind the wheel performance and oculomotor correlates. *Human Factors: The Journal of the Human Factors and Ergonomics Society*, 53(1), 13–21.

He, J., Ellis, J., Choi, W., & Wang, P. (2015). Driving while reading using Google glass versus using a smart phone: Which is more distracting to driving performance? In *Proceedings of the Eighth International Driving Symposium on Human Factors in Driver Assessment, Training and Vehicle Design*, 275–281, Salt Lake City, UT.

Healey, J. A., & Picard, R. W. (2005). Detecting stress during real-world driving tasks using physiological sensors. *IEEE Transactions on Intelligent Transportation Systems*, 6(2), 156–166.

Health and Safety at Work etc. Act (1974). www.legislation.gov.uk. Accessed 21/07/2016.

Hedlund, J., Simpson, H., & Mayhew, D. (2005). *International Conference on Distracting Driving: Summary of Proceedings and Recommendations*. Cited in: Regan, M. A., Hallett, C., & Gordon, C. P. (2011). Driver distraction and driver inattention: Definition, relationship and taxonomy. *Accident Analysis & Prevention*, 43(5), 1771–1781.

Hettinger, L. J., Kirlik, A., Goh, Y. M., & Buckle, P. (2015). Modelling and simulation of complex sociotechnical systems: Envisioning and analysing work environments. *Ergonomics*, 58(4), 600–614.

Hickman, J. S., & Hanowski, R. J. (2012). An assessment of commercial motor vehicle driver distraction using naturalistic driving data. *Traffic Injury Prevention*, 13(6), 612–619.

Hignett, S. (2005). Qualitative methodology in ergonomics. In Wilson, J. R., & Megaw, E. (Eds.) *Evaluation of Human Work*. Taylor & Francis, London.

Hignett, S., Jones, E. L., Miller, D., Wolf, L., Modi, C., Shahzad, M. W., Buckle, P., Banerjee, J., & Catchpole, K. (2015). Human factors and ergonomics and quality improvement science: Integrating approaches for safety in healthcare. *BMJ Quality & Safety*, 24(4), 250–254.

Hockey, G. R. J. (1997). Compensatory control in the regulation of human performance under stress and high work-load: A cognitive–energetical framework. *Biological Psychology*, 45, 73–93.

Hoel, J., Jaffard, M., & Van Elslande, P. (2010). Attentional competition between tasks and its implications. Cited in: Regan, M. A., Hallett, C., & Gordon, C. P. (2011). Driver distraction and driver inattention: Definition, relationship and taxonomy. *Accident Analysis & Prevention*, 43(5), 1771–1781.

Hollnagel, E. (2009). *The ETTO Principle: Efficiency-Thoroughness Trade-Off: Why Things That Go Right Sometimes Go Wrong*. Ashgate Publishing, Ltd., Farnham, UK.

Hollnagel, E., Woods, D. D., & Leveson, N. (2007). *Resilience Engineering: Concepts and Precepts*. Ashgate Publishing, Ltd., Aldershot, UK.

Holtzblatt, K., & Jones, S. (1993). Contextual inquiry: A participatory technique for system design. In Schuler, D., & Aki, N. (Eds.) *Participatory Design: Principles and Practices*. Lawrence Erlbaum Associated Publishers, Hillsdale, NJ, 177–210.

Horberry, T., Anderson, J., Regan, M. A., Triggs, T. J., & Brown, J. (2006). Driver distraction: The effects of concurrent in-vehicle tasks, road environment complexity and age on driving performance. *Accident Analysis & Prevention*, 38(1), 185–191.

Horrey, W. J., & Lesch, M. F. (2008). Factors related to drivers' self-reported willingness to engage in distracting in-vehicle activities. In *Proceedings of the Human Factors and Ergonomics Society Annual Meeting,* 52(19), 1546–1550, Sage, Los Angeles, CA.

Horrey, W. J., & Lesch, M. F. (2009). Driver-initiated distractions: Examining strategic adaptation for in-vehicle task initiation. *Accident Analysis & Prevention*, 41(1), 115–122.

Horrey, W. J., Lesch, M. F., & Garabet, A. (2008). Assessing the awareness of performance decrements in distracted drivers. *Accident Analysis & Prevention*, 40(2), 675–682.

Horrey, W. J., Lesch, M. F., Garabet, A., Simmons, L., & Maikala, R. (2017). Distraction and task engagement: How interesting and boring information impact driving performance and subjective and physiological responses. *Applied Ergonomics*, 58, 342–348.

Horrey, W. J., & Simons, D. (2007). Examining cognitive interference and adaptive safety behaviours in tactical vehicle control. *Ergonomics*, 50(8), 1340–1350.

Horrey, W. J., & Wickens, C. D. (2004). Driving and side task performance: The effects of display clutter, separation, and modality. *Human Factors: The Journal of the Human Factors and Ergonomics Society*, 46(4), 611–624.

Horrey, W. J., & Wickens, C. D. (2006). Examining the impact of cell phone conversations on driving using meta-analytic techniques. *Human Factors: The Journal of the Human Factors and Ergonomics Society*, 48(1), 196–205.

Horrey, W. J., Wickens, C. D., & Consalus, K. P. (2006). Modeling drivers' visual attention allocation while interacting with in-vehicle technologies. *Journal of Experimental Psychology: Applied*, 12(2), 67.

Hosking, S. G., Young, K. L., & Regan, M. A. (2009). The effects of text messaging on young drivers. *Human Factors: The Journal of the Human Factors and Ergonomics Society*, 51(4), 582–592.

Huemer, A. K., & Vollrath, M. (2011). Driver secondary tasks in Germany: Using interviews to estimate prevalence. *Accident Analysis & Prevention*, 43(5), 1703–1712.

Hughes, B. P., Newstead, S., Anund, A., Shu, C. C., & Falkmer, T. (2015). A review of models relevant to road safety. *Accident Analysis & Prevention*, 74, 250–270.

Jahn, G., Oehme, A., Krems, J. F., & Gelau, C. (2005). Peripheral detection as a workload measure in driving: Effects of traffic complexity and route guidance system use in a driving study. *Transportation Research Part F: Traffic Psychology and Behaviour*, 8(3), 255–275.

Jamson, A. H., Merat, N., Carsten, O. M., & Lai, F. C. (2013). Behavioural changes in drivers experiencing highly-automated vehicle control in varying traffic conditions. *Transportation Research Part C: Emerging Technologies*, 30, 116–125.

References

Jamson, A. H., Westerman, S. J., Hockey, G. R. J., & Carsten, O. M. (2004). Speech-based e-mail and driver behavior: Effects of an in-vehicle message system interface. *Human Factors: The Journal of the Human Factors and Ergonomics Society*, 46(4), 625–639.

Janitzek, T., Brenck, A., Jamson, S., Carsten, O. M., & Eksler, V. (2010). Study on the regulatory situation in the member states regarding brought-in (i.e. nomadic) devices and their use in vehicles. *Study tendered by the European Commission (SMART 2009/0065) Final Report.*

Japan Automobile Manufacturers Association (2004). Guideline for In-Vehicle Display Systems, Version 3.0, Tokyo, Japan.

Jensen, B. S., Skov, M. B., & Thiruravichandran, N. (2010). Studying driver attention and behaviour for three configurations of GPS navigation in real traffic driving. In *Proceedings of the SIGCHI Conference on Human Factors in Computing Systems*, 1271–1280, ACM, Atlanta, GA.

Johansson, R. (2009). Vision Zero–Implementing a policy for traffic safety. *Safety Science*, 47(6), 826–831.

Kahneman, D., Slovic, P., & Tversky, A. (1982). *Judgment Under Uncertainty: Heuristics and Biases.* Cambridge University Press, Cambridge, NY.

Kaptein, N., Theeuwes, J., & Van Der Horst, R. (1996). Driving simulator validity: Some considerations. *Transportation Research Record: Journal of the Transportation Research Board*, 1550, 30–36.

Kass, S. J., Cole, K. S., & Stanny, C. J. (2007). Effects of distraction and experience on situation awareness and simulated driving. *Transportation Research Part F: Traffic Psychology and Behaviour*, 10(4), 321–329.

Kazaras, K., Kirytopoulos, K., & Rentizelas, A. (2012). Introducing the STAMP method in road tunnel safety assessment. *Safety Science*, 50(9), 1806–1817.

Kervick, A. A., Hogan, M. J., O'Hora, D., & Sarma, K. M. (2015). Testing a structural model of young driver willingness to uptake Smartphone Driver Support Systems. *Accident Analysis & Prevention*, 83, 171–181.

King, J. L., & Lyytinen, K. (2005). Automotive informatics: Information technology and enterprise transformation in the automobile industry. In Dutton, W. H., Kahin, B., O'Callaghan, R., & Wychoff, A. W. (Eds.) *Transforming Enterprise: The Economic and Social Implications of Information Technology.* MIT Press, Cambridge, MA, 283–312.

Kircher, A., Vogel, K., Tornos, J., Bolling, A., Nilsson, L., Patten, C., Malmstrom, T., & Ceco, C. (2004). *Mobile Telephone Simulator Study.* Swedish National Road and Transport Research Institute, Linköping.

Kircher, K., & Ahlstrom, C. (2010). Predicting visual distraction using driving performance data. *Annals of Advances in Automotive Medicine*, 54, 333–342.

Kircher, K., & Ahlstrom, C. (2016). Minimum required attention: A human-centred approach to driver inattention. *Human Factors: The Journal of the Human Factors and Ergonomics Society*, 59(3), 471–484.

Kircher, K., & Ahlstrom, C. (2018). Evaluation of methods for the assessment of attention while driving. *Accident Analysis & Prevention*, 144, 40–47.

Kircher, K., Ahlström, C., Gregersen, N. P., & Patten, C. (2013). Why Sweden should not do as everybody else does. In *3rd International Conference on Driver Distraction and Inattention*, September, Gothenburg, Sweden, (No. 13–P).

Kircher, K., Fors, C., & Ahlstrom, C. (2014). Continuous versus intermittent presentation of visual eco-driving advice. *Transportation Research Part F: Traffic Psychology and Behaviour*, 24, 27–38.

Kircher, K., Patten, C., & Ahlström, C. (2011). Mobile telephones and other communication devices and their impact on traffic safety—A review of the literature. *Report 729A*, Swedish National Road and Transport Research Institute, Linköping, Sweden.

Klauer, S. G., Dingus, T. A., Neale, V. L., Sudweeks, J. D., & Ramsey, D. J. (2006). The impact of driver inattention on near-crash/crash risk: An analysis using the 100-car naturalistic driving study data. *Report: DOT HS 810 594*, National Highway Traffic Safety Administration, Washington, DC.

Klein, G. A. (1989). Recognition-primed decisions. In Rouse, W. (Ed.) *Advances in Man-Machine Systems Research*. JAI Press, Greenwich, CT, 47–92.

Klein, G. (2008). Naturalistic decision making. *Human Factors: The Journal of the Human Factors and Ergonomics Society*, 50(3), 456–460.

Klein, G. A., Calderwood, R., & Clinton-Cirocco, A. (1986). Rapid decision making on the fire ground. In *Proceedings of the Human Factors Society Annual Meeting*, 30(6), 576–580, Sage, Los Angeles, CA.

Konstantopoulos, P., Chapman, P., & Crundall, D. (2010). Driver's visual attention as a function of driving experience and visibility: Using a driving simulator to explore drivers' eye movements in day, night and rain driving. *Accident Analysis & Prevention*, 42(3), 827–834.

Korpinena, L., & Pääkkçnen, R. (2012). Accidents and close call situations connected to the use of mobile phones. *Accident Analysis and Prevention*, 45, 75–82.

Kountouriotis, G. K., & Merat, N. (2016). Leading to distraction: Driver distraction, lead car, and road environment. *Accident Analysis & Prevention*, 89, 22–30.

Kurasaki, K. S. (2000). Intercoder reliability for validating conclusions drawn from open-ended interview data. *Field Methods*, 12(3), 179–194.

Kutila, M., Jokela, M., Markkula, G., & Rué, M. R. (2007). Driver distraction detection with a camera vision system. In *IEEE International Conference on Image Processing*, 6, (VI-201), October, San Antoniom, TX.

Lajunen, T., & Summala, H. (2003). Can we trust self-reports of driving? Effects of impression management on driver behaviour questionnaire responses. *Transportation Research Part F: Traffic Psychology and Behaviour*, 6(2), 97–107.

Lansdown, T. C. (2012). Individual differences and propensity to engage with in-vehicle distractions–A self-report survey. *Transportation Research Part F: Traffic Psychology and Behaviour*, 15(1), 1–8.

Lansdown, T. C., Brook-Carter, N., & Kersloot, T. (2004). Distraction from multiple in-vehicle secondary tasks: Vehicle performance and mental workload implications. *Ergonomics*, 47(1), 91–104.

Lansdown, T. C., Stephens, A. N., & Walker, G. H. (2015). Multiple driver distractions: A systemic transport problem. *Accident Analysis & Prevention*, 74, 360–367.

Lam, L. T. (2002). Distractions and the risk of car crash injury: The effect of drivers' age. *Journal of Safety Research*, 33(3), 411–419.

Lamble, D., Kauranen, T., Laakso, M., & Summala, H. (1999). Cognitive load and detection thresholds in car following situations: Safety implications for using mobile (cellular) telephones while driving. *Accident Analysis & Prevention*, 31(6), 617–623.

Lamble, D., Sirpa R., & Heikki S. (2002) Mobile phone use while driving: Public opinions on restrictions. *Transportation*, 29(3), 223–236.

Larsson, P., Dekker, S. W., & Tingvall, C. (2010). The need for a systems theory approach to road safety. *Safety Science*, 48(9), 1167–1174.

Layder, D. (1998). *Sociological Practice: Linking Theory and Social Research*. Sage, London.

Lee, C., Mehler, B., Reimer, B., & Coughlin, J. F. (2015). User perceptions toward in-vehicle technologies: Relationships to age, health, preconceptions, and hands-on experience. *International Journal of Human-Computer Interaction*, 31(10), 667–681.

Lee, J. D. (2014). Dynamics of driver distraction: The process of engaging and disengaging. *Annals of Advances in Automotive Medicine*, 58, 24–32.

References

Lee, J. D., Caven, B., Haake, S., & Brown, T. L. (2001). Speech-based interaction with in-vehicle computers: The effect of speech-based e-mail on drivers' attention to the roadway. *Human Factors: The Journal of the Human Factors and Ergonomics Society*, 43(4), 631–640.

Lee, J. D., Hoffman, J. D., & Hayes, E. (2004). Collision warning design to mitigate driver distraction. In *Proceedings of the SIGCHI Conference on Human Factors in Computing Systems*, 65–72, ACM, New York, NY.

Lee, J. D., Roberts, S. C., Hoffman, J. D., & Angell, L. S. (2012). Scrolling and driving: How an MP3 player and its aftermarket controller affect driving performance and visual behavior. *Human Factors: The Journal of the Human Factors and Ergonomics Society*, 54(2), 250–263.

Lee, J. D., & Strayer, D. L. (2004). Preface to the special section on driver distraction. *Human Factors: The Journal of the Human Factors and Ergonomics Society*, 46(4), 583–586.

Lee, J. D., Young, K. L., & Regan, M. A. (2008) Defining driver distraction. In Regan, M. A., Lee, J. D., & Young, K. (Eds.) *Driver Distraction: Theory, Effects, and Mitigation*, CRC Press, Boca Raton, FL.

Lerner, N. (2005). Deciding to be distracted. In *Proceedings of the Third International Driving Symposium on Human Factors in Driver Assessment, Training and Vehicle Design*, 499–505, Rockport, ME.

Lerner, N., & Boyd, S. (2005). Task report: On-road study of willingness to engage in distraction tasks *Report: DOT HS 809 863*, National Highway Traffic Safety Administration, Washington, DC.

Lerner, N., Singer, J., & Huey, R. (2008). Driver strategies for engaging in distracting tasks using in-vehicle technologies. *Report No: DOT HS 810 919*, US Department of Transportation, Washington, DC.

Leveson, N. G. (2004). A new accident model for engineering safer systems. *Safety Science*, 42(4), 237–270.

Leveson, N. G. (2011). Applying systems thinking to analyze and learn from events. *Safety Science*, 49(1), 55–64.

Leveson, N. G. (2012.) *Engineering a Safer World. Systems Thinking Applied to Safety*. MIT Press, Cambridge.

Liang, Y., & Lee, J. D. (2010). Combining cognitive and visual distraction: Less than the sum of its parts. *Accident Analysis & Prevention*, 42(3), 881–890.

Liang, Y., Reyes, M. L., & Lee, J. D. (2007). Real-time detection of driver cognitive distraction using support vector machines. *IEEE Transactions on Intelligent Transportation Systems*, 8(2), 340–350.

Liu, Y. C. (2003). Effects of using head-up display in automobile context on attention demand and driving performance. *Displays*, 24(4), 157–165.

Loukopoulos, L. D., Dismukes, R. K., & Barshi, I. (2001). Cockpit interruptions and distractions: A line observation study. In *Proceedings of the 11th International Symposium on Aviation Psychology*, 1–6, Ohio State University Press, Columbus.

Ma, R., & Kaber, D. B. (2005). Situation awareness and workload in driving while using adaptive cruise control and a cell phone. *International Journal of Industrial Ergonomics*, 35(10), 939–953.

Manser, M. P., Ward, N. J., Kuge, N., & Boer, E. R. (2004). Influence of a driver support system on situation awareness and information processing in response to lead vehicle braking. In *Proceedings of the Human Factors and Ergonomics Society Annual Meeting*, 48(19), 2359–2363, Sage, Los Angeles, CA.

Marshall, M. N. (1996). Sampling for qualitative research. *Family Practice*, 13(6), 522–526.

Matthews, B. W. (1975). Comparison of the predicted and observed secondary structure of T4 phage lysozyme. *Biochimica et Biophysica Acta (BBA)-Protein Structure*, 405(2), 442–451.

Matthews, R., Legg, S., & Charlton, S. (2003). The effect of cell phone type on drivers' subjective workload during concurrent driving and conversing. *Accident Analysis and Prevention*, 35, 441–450.

McCartt, A. T., & Geary, L. L. (2004). Longer term effects of New York State's law on drivers' handheld cell phone use. *Injury Prevention*, 10(1), 11–15.

McCartt, A. T., Hellinga, L. A., & Bratiman, K. A. (2006). Cell phones and driving: Review of research. *Traffic Injury Prevention*, 7(2), 89–106.

McEvoy, S. P., Stevenson, M. R., & Woodward, M. (2006). The impact of driver distraction on road safety: Results from a representative survey in two Australian states. *Injury Prevention*, 12(4), 242–247.

McEvoy, S. P., Stevenson, M. R., & Woodward, M. (2007). The prevalence of, and factors associated with, serious crashes involving a distracting activity. *Accident Analysis & Prevention*, 39(3), 475–482.

Mehlenbacher, B., Wogalter, M. S., & Laughery, K. R. (2002). On the reading of product owner's manuals: Perceptions and product complexity. In *Proceedings of the Human Factors and Ergonomics Society Annual Meeting*, 46(6), 730–734.

Meng, L. (2004). About egocentric geovisualisation. In *Proceedings of the 12th International Conference on Geoinformatics: Bridging the Pacific and Atlantic*, June 7–14, University of Gavle, Sweden.

Merat, N., Jamson, A. H., Lai, F. C., & Carsten, O. (2012). Highly automated driving, secondary task performance, and driver state. *Human Factors*, 54(5), 762–771.

Merat, N., Jamson, A. H., Lai, F. C., Daly, M., & Carsten, O. M. (2014). Transition to manual: Driver behaviour when resuming control from a highly automated vehicle. *Transportation Research Part F: Traffic Psychology and Behaviour*, 27, 274–282.

Metz, B., Landau, A., & Hargutt, V. (2015). Frequency and impact of hands-free telephoning while driving—Results from naturalistic driving data. *Transportation Research Part F: Traffic Psychology and Behaviour*, 29(0), 1–13.

Metz, B., Schömig, N., & Krüger, H. P. (2011). Attention during visual secondary tasks in driving: Adaptation to the demands of the driving task. *Transportation Research Part F: Traffic Psychology and Behaviour*, 14(5), 369–380.

Michon, J. A. (1985). A critical view of driver behavior models: What do we know, what should we do. In Evans, L., & Schwing, R.C. (Eds.) *Human Behavior and Traffic Safety*. Plenum Press, New York, NY, 485–520.

Miller, D., Sun, A., Johns, M., Ive, H., Sirkin, D., Aich, S., & Ju, W. (2015). Distraction becomes engagement in automated driving. In *Proceedings of the Human Factors and Ergonomics Society Annual Meeting*, 59(1), 1676–1680, Sage, Los Angeles, CA.

Mitsopoulos-Rubens, E., Trotter, M. J., & Lenné, M. G. (2011). Effects on driving performance of interacting with an in-vehicle music player: A comparison of three interface layout concepts for information presentation. *Applied Ergonomics*, 42(4), 583–591.

Mitzner, T. L., Boron, J. B., Fausset, C. B., Adams, A. E., Charness, N., Czaja, S. J., Dijkstra, K., Fisk, A. D., Rogers, W. A., & Sharit, J. (2010). Older adults talk technology: Technology usage and attitudes. *Computers in Human Behavior*, 26(6), 1710–1721.

Mizenko, A. J., Tefft, B. C., Arnold, L. S., & Grabowski, J. G. (2015). The relationship between age and driving attitudes and behaviors among older Americans. *Injury Epidemiology*, 2(1), 1–10.

Monk, C. A., Boehm-Davis, D. A., Mason, G., & Trafton, J. G. (2004). Recovering from interruptions: Implications for driver distraction research. *Human Factors: The Journal of the Human Factors and Ergonomics Society*, 46(4), 650–663.

References

Moray, N. (1999). The psychodynamics of human-machine interaction. In Harris, D. (Ed.) *Engineering Psychology and Cognitive Ergonomics*. Ashgate, Aldershot, 4, 225–235.

Musselwhite, C. B., & Haddad, H. (2010). Exploring older drivers' perceptions of driving. *European Journal of Ageing*, 7(3), 181–188.

Naweed, A. (2013). Psychological factors for driver distraction and inattention in the Australian and New Zealand rail industry. *Accident Analysis & Prevention*, 60, 193–204.

Neale, V. L., Dingus, T. A., Klauer, S. G., Sudweeks, J., & Goodman, M. (2002). An overview of the 100-car naturalistic study and findings. In *Proceedings of the 19th International Technical Conference on the Enhanced Safety of Vehicles (CD–ROM)*, National Highway Traffic Safety Administration, Washington, DC. Paper No. 05–0400.

Neale, V. L., Dingus, T. A., Klauer, S. G., Sudweeks, J., & Goodman, M. (2005). *An overview of the 100-car naturalistic study and findings*. National Highway Traffic Safety Administration, Paper 05–0400. USA.

Nelson, E., Atchley, P., & Little, T. D. (2009). The effects of perception of risk and importance of answering and initiating a cellular phone call while driving. *Accident Analysis & Prevention*, 41(3), 438–444.

Nemme, H. E., & White, K. M. (2010). Texting while driving: Psychosocial influences on young people's texting intentions and behaviour. *Accident Analysis & Prevention*, 42(4), 1257–1265.

Newnam, S., & Goode, N. (2015). Do not blame the driver: A systems analysis of the causes of road freight crashes. *Accident Analysis & Prevention*, 76, 141–151.

NHTSA (2013). *Visual-manual NHTSA driver distraction guidelines for in-vehicle electronic devices*. National Highway Traffic Safety Administration (NHTSA), Department of Transportation (DOT), Washington, DC.

NHTSA (2016). Distracted Driving 2014. Summary of Statistical Findings. DOT HS 812 260. https://crashstats.nhtsa.dot.gov/Api/Public/ViewPublication/812260. Accessed 26/10/2016.

NHTSA (2017). Automated Driving Systems 2.0: A Vision for Safety. US Department of Transportation. www.nhtsa.gov/sites/nhtsa.dot.gov/files/documents/13069a-ads2.0_090617_v9a_tag.pdf. Accessed 01/02/2018.

Noy, Y. I. (1989). Intelligent route guidance: Will the new horse be as good as the old? In *Vehicle Navigation and Information Systems Conference, Conference Record*. IEEE, Toronto. 49–55.

O'Cathain, A., & Thomas, K. J. (2004). "Any other comments?" Open questions on questionnaires—A bane or a bonus to research? *BMC Medical Research Methodology*, 4(1), 25.

Olson, R. L., Hanowski, R. J., Hickman, J. S., & Bocanegra, J. (2009). *Driver distraction in commercial vehicle operations* (No. FMCSA-RRT-09–042). Federal Motor Carrier Safety Administration, Washington, DC.

Olsen, E. C. B., Lerner, N., Perel, M., & Simons-Morton, B. G. (2005). In-car electronic device use among teen drivers. In *Transportation Research Board Meeting*, Washington, DC.

Olsen, N. S., & Shorrock, S. T. (2010). Evaluation of the HFACS-ADF safety classification system: Inter-coder consensus and intra-coder consistency. *Accident Analysis & Prevention*, 42(2), 437–444.

Office for National Statistics (2013). *Crime in England and Wales, period ending September 2013*. Released 23 January 2014. www.ons.gov.uk/ons/rel/crime-stats/crime-statistics/index.html.

Papadakos, P. J. (2013). The rise of electronic distraction in health care is addiction to devices contributing. *Journal of Aesthetic Clinical Research*, 4(3), e112.

Parnell, K. J., Stanton, N. A., & Plant, K. L. (2016). Exploring the mechanisms of distraction from in-vehicle technology: The development of the PARRC model. *Safety Science*, 87, 25–37.

Parnell, K. J., Stanton, N. A., & Plant, K. L. (2017a). What's the law got to do with it? Legislation regarding in-vehicle technology use and its impact on driver distraction. *Accident Analysis & Prevention*, 100, 1–14.

Parnell, K. J., Stanton, N. A., & Plant, K. (2017b). Where are we on driver distraction? Methods, approaches and recommendations. *Theoretical Issues in Ergonomics Science*, 19, 1–28.

Parnell, K. J., Stanton, N. A., & Plant, K. L. (2018a). Creating the environment for driver distraction: A thematic framework of sociotechnical factors. *Applied Ergonomics*, 68, 213–228.

Parnell, K. J., Stanton, N. A., & Plant, K. L. (2018b). What technologies do people engage with while driving and why? *Accident Analysis & Prevention*, 111, 222–237.

Parnell, K. J., Stanton, N. A., & Plant, K. L. (2018c). Good intentions: Drivers' decisions to engage with technology on the road and in a driving simulator. *Cognition, Technology and Work*. doi:10.1007/s10111-018-0504-0.

Patel, J., Ball, D. J., & Jones, H. (2008). Factors influencing subjective ranking of driver distractions. *Accident Analysis & Prevention*, 40(1), 392–395.

Patten, C. J., Kircher, A., Östlund, J., & Nilsson, L. (2004). Using mobile telephones: Cognitive workload and attention resource allocation. *Accident Analysis & Prevention*, 36(3), 341–350.

Patten, C. J., Kircher, A., Östlund, J., Nilsson, L., & Svenson, O. (2006). Driver experience and cognitive workload in different traffic environments. *Accident Analysis & Prevention*, 38(5), 887–894.

Patton, M. Q. (1990). *Qualitative Evaluation and Research Methods*. Sage, Thousand Oaks, CA.

Pedic, F., & Ezrakhovich, A. (1999). A literature review: The content characteristics of effective VMS. *Road & Transport Research*, 8(2), 3.

Perrow, C. (1984). *Normal Accidents: Living with High-Risk Technologies*. Basic Books, New York.

Pettitt, M. A., Burnett, G., & Stevens, A. (2005). Defining driver distraction. In: *Proceedings of the 12th ITS World Congress*. ITS America, San Francisco.

Pettitt, M. A., Burnett, G. E., Bayer, S., & Stevens, A. (2006). Assessment of the occlusion technique as a means for evaluating the distraction potential of driver support systems. *IEE Proceedings in Intelligent Transport Systems*, 4(1), 259–266.

Phillips, J. K., Klein, G., & Sieck, W. R. (2004). Expertise in judgment and decision making: A case for training intuitive decision skills. In Koehler, D. J., & Harvey, N. (Eds.) *Blackwell Handbook of Judgment and Decision Making*. Blackwell, Oxford, 297–315.

Plant, K. L., & Stanton, N. A. (2012). Why did the pilots shut down the wrong engine? Explaining errors in context using Schema Theory and the Perceptual Cycle Model. *Safety Science*, 50(2), 300–315.

Plant, K. L., & Stanton, N. A. (2013). What is on your mind? Using the perceptual cycle model and critical decision method to understand the decision-making process in the cockpit. *Ergonomics*, 56(8), 1232–1250.

Plant, K. L., & Stanton, N. A. (2016). *Distributed Cognition and Reality: How Pilots and Crews Make Decisions*. CRC Press, Baco Raton, FL.

Pope, C. N., Bell, T. R., & Stavrinos, D. (2017). Mechanisms behind distracted driving behavior: The role of age and executive function in the engagement of distracted driving. *Accident Analysis & Prevention*, 98, 123–129.

Postman, N. (1993). *Technopoly: The Surrender of Culture to Technology*. Vintage Books, New York.

Pöysti, L., Rajalin, S., & Summala, H. (2005). Factors influencing the use of cellular (mobile) phone during driving and hazards while using it. *Accident Analysis & Prevention*, 37(1), 47–51.

References

RAC (2013). RAC Motoring Report 2013. RAC Foundation for Motoring and the Environment. www.rac.co.uk/RAC/files/5b/5bfb0586-d8d9-41f6-b8a2-896b45b38cb3.pdf. Accessed 10/06/2017.

RAC (2016). RAC Report on Motoring 2016. The Road to the Future. https://www.rac.co.uk/pdfs/report-on-motoring/rac-report-on-motoring-2016-outline.pdf. Accessed 28/04/2016.

Rafferty, L. A., Stanton, N. A., & Walker, G. H. (2010). The famous five factors in teamwork: A case study of fratricide. *Ergonomics,* 53(10), 1187–1204.

Rakauskas, M. E., Gugerty, L. J., & Ward, N. J. (2004). Effects of naturalistic cell phone conversations on driving performance. *Journal of Safety Research*, 35(4), 453–464.

Ranney, T. A., Harbluk, J. L., & Noy, Y. I. (2005). Effects of voice technology on test track driving performance: Implications for driver distraction. *Human Factors: The Journal of the Human Factors and Ergonomics Society*, 47(2), 439–454.

Ranney, T. A., Mazzae, E., Garrott, R., & Goodman, M. J. (2000). *NHTSA driver distraction research: Past, present, and future.* Transportation Research Centre Inc., East Liberty, OH.

Rasmussen, J. (1997). Risk management in a dynamic society: A modelling problem. *Safety Science*, 27(2), 183–213.

Reason, J. (1990). *Human Error.* Cambridge University Press, Cambridge, UK.

Reason, J. (1995). Understanding adverse events: Human factors. *Quality in Health Care*, 4(2), 80–89.

Reason, J. (2000). Human error: Models and management. *BMJ*, 320(7237), 768–770.

Reason, J., Manstead, A., Stradling, S., Baxter, J., & Campbell, K. (1990). Errors and violations on the roads: A real distinction? *Ergonomics*, 33(10), 1315–1332.

Recarte, M. A., & Nunes, L. M. (2000). Effects of verbal and spatial-imagery tasks on eye fixations while driving. *Journal of Experimental Psychology: Applied*, 6(1), 31.

Redelmeier, D. A., & Tibshirani, R. J. (1997). Association between cellular-telephone calls and motor vehicle collisions. *New England Journal of Medicine*, 336(7), 453–458.

Reed, M. P., & Green, P. A. (1999). Comparison of driving performance on-road and in a low-cost simulator using a concurrent telephone dialling task. *Ergonomics*, 42(8), 1015–1037.

Regan, M. A., Lee, J. D., & Young, K. (Eds.) (2008). *Driver Distraction: Theory, Effects, and Mitigation.* CRC Press, Boca Raton, FL.

Regan, M. A., Hallett, C., & Gordon, C. P. (2011). Driver distraction and driver inattention: Definition, relationship and taxonomy. *Accident Analysis & Prevention*, 43(5), 1771–178.

Regan, M. A., Young, K. L., Lee, J. D., & Gordon, C. P. (2009). Sources of driver distraction. In Regan, M. A., Lee, J. D., & Young, K. (Eds.), *Driver Distraction: Theory, Effects, and Mitigation.* CRC Press, Boca Raton, FL.

Reimer, B. (2009). Impact of cognitive task complexity on drivers' visual tunnelling. *Transportation Research Record: Journal of the Transportation Research Board*, 2138, 13–19.

Reinach, S., & Viale, A. (2006). Application of a human error framework to conduct train accident/incident investigations. *Accident Analysis & Prevention*, 38(2), 396–406.

Richards, L., & Richards, T. (1991). The transformation of qualitative method: Computational paradigms and research processes. In Fielding, N. G., & Lee, R. M. (Eds.) *Using Computers in Qualitative Research.* Sage, London, UK 38–53.

Ritchie, J., & Lewis, J. (2003). *Qualitative Research Practice.* Sage, London, UK.

Road Traffic Act (1988). Section 2, 41D, inserted (27.2.2007) by Road Traffic Act 2006.

Road Safety Act (2006).www.legislation.gov.uk/ukpga/2006/49/section/20. Accessed 21/07/2016.

Robert, G., & Hockey, J. (1997). Compensatory control in the regulation of human performance under stress and high workload: A cognitive-energetical framework. *Biological Psychology*, 45(1), 73–93.

Rosencrantz, H., Edvardsson, K., & Hansson, S. O. (2007). Vision zero–Is it irrational? *Transportation Research Part A: Policy and Practice*, 41(6), 559–567.

Rouzikhah, H., King, M., & Rakotonirainy, A. (2013). Examining the effects of an eco-driving message on driver distraction. *Accident Analysis & Prevention*, 50, 975–983.

Rudin-Brown, C. M., & Parker, H. A. (2004). Behavioural adaptation to adaptive cruise control (ACC): Implications for preventive strategies. *Transportation Research Part F: Traffic Psychology and Behaviour*, 7(2), 59–76.

Rudin-Brown, C. M., Parker, H. A., & Malisia, A. R. (2003). Behavioural adaptation to adaptive cruise control. *In Proceedings of the 47th Annual Meeting of the Human Factors and Ergonomics Society*, October 13–17, Denver, CO.

Sabey, B. E., & Taylor, H. (1980). The known risks we run: The highway. In Schwing, R. C., & Albers, W. A. (Eds.) *Societal Risk Assessment*. Springer, Boston, MA 43–70.

SAE On-road Automated Vehicles Standards Committee (2016). *Taxonomy and Definitions for Terms Related to On-Road Motor Vehicle Automated Driving Systems*. Accessible: http://standards.sae.org/j3016_201401/.

Salmon, P. M., Cornelissen, M., & Trotter, M. J. (2012). Systems-based accident analysis methods: A comparison of Accimap, HFACS, and STAMP. *Safety Science*, 50(4), 1158–1170.

Salmon, P. M., Goode, N., Spiertz, A., Thomas, M., Grant, E., & Clacy, A. (2017). Is it really good to talk? Testing the impact of providing concurrent verbal protocols on driving performance. *Ergonomics*, 60(6), 770–779.

Salmon, P. M., Goode, N., Stevens, E., Walker, G., & Stanton, N. A. (2015). The elephant in the room: Normal performance and accident analysis. In *International Conference on Engineering Psychology and Cognitive Ergonomics,* 275–285, Springer, Cham.

Salmon, P. M., Goode, N., Taylor, N., Lenné, M. G., Dallat, C. E., & Finch, C. F. (2017). Rasmussen's legacy in the great outdoors: A new incident reporting and learning system for led outdoor activities. *Applied Ergonomics*, 59, 637–648.

Salmon, P. M., Lenné, M. G., Stanton, N. A., Jenkins, D. P., & Walker, G. H. (2010). Managing error on the open road: The contribution of human error models and methods. *Safety Science*, 48(10), 1225–1235.

Salmon, P. M., McClure, R., & Stanton, N. A. (2012). Road transport in drift? Applying contemporary systems thinking to road safety. *Safety Science*, 50(9), 1829–1838.

Salmon, P. M., Read, G. J., & Stevens, N. J. (2016). Who is in control of road safety? A STAMP control structure analysis of the road transport system in Queensland, Australia. *Accident Analysis & Prevention*, 96, 140–151.

Salmon, P. M., Stanton, N. A., Lenné, M., Jenkins, A. P., Rafferty, L., & Walker, G. H. (2011). *Human Factors Methods and Accident Analysis: Practical Guidance and Case Study Applications*. Ashgate Publishing, Ltd., Aldershot, UK.

Salmon, P. M., Stanton, N. A., & Young, K. L. (2012). Situation awareness on the road: Review, theoretical and methodological issues, and future directions. *Theoretical Issues in Ergonomics Science*, 13(4), 472–492.

Salmon, P. M., Walker, G. H., Read, G. J. M., Goode, N., & Stanton, N. A. (2017). Fitting methods to paradigms: Are ergonomics methods fit for systems thinking? *Ergonomics*, 60(2), 194–205.

Salmon, P. M., Young, K. L., & Regan, M. A. (2011). Distraction 'on the buses': A novel framework of ergonomics methods for identifying sources and effects of bus driver distraction. *Applied Ergonomics*, 42(4), 602–610.

Salvucci, D. D., & Taatgen, N. A. (2008). Threaded cognition: An integrated theory of concurrent multitasking. *Psychological Review*, 115(1), 101–130.

Sanders, M. S., & McCormick, E. J. (1994). *Human Factors in Engineering and Design*. McGraw-Hill, New York, NY.

References

Sawyer, B. D., Finomore, V. S., Calvo, A. A., & Hancock, P. A. (2014). Google Glass: A driver distraction cause or cure? *Human Factors: The Journal of the Human Factors and Ergonomics Society*, 56(7), 1307–1321.

Scarborough, A., & Pounds, J. (2001). Retrospective human factors analysis of ATC operational errors. In *11th International Symposium on Aviation Psychology*, 1–5, Columbus, OH.

Schömig, N., & Metz, B. (2013). Three levels of situation awareness in driving with secondary tasks. *Safety Science*, 56, 44–51.

Scott-Parker, B., Goode, N., & Salmon, P. (2015). The driver, the road, the rules... and the rest? A systems-based approach to young driver road safety. *Accident Analysis & Prevention*, 74, 297–305.

Schwebel, D. C., Stavrinos, D., Byington, K. W., Davis, T., O'Neal, E. E., & De Jong, D. (2012). Distraction and pedestrian safety: How talking on the phone, texting, and listening to music impact crossing the street. *Accident Analysis & Prevention*, 45, 266–271.

Shappell, S. A., & Wiegmann, D. A. (2003). A human error analysis of general aviation controlled flight into terrain accidents occurring between 1990–1998 *Report No: DOT/FAA/AM-03/4*, Office of Aerospace Medicine, Washington, DC.

Shappell, S. A., & Wiegmann, D. A. (2012). *A Human Error Approach to Aviation Accident Analysis: The Human Factors Analysis and Classification System*. Ashgate Publishing, Ltd., Aldershot, UK.

Sharples, S., Shalloe, S., Burnett, G., & Crundall, D. (2016). Journey decision making: The influence on drivers of dynamic information presented on variable message signs. *Cognition, Technology & Work*, 18(2), 303–317.

Sheridan, T. B. (2004). Driver distraction from a control theory perspective. *Human Factors: The Journal of the Human Factors and Ergonomics Society*, 46(4), 587–599.

Shinar, D., Tractinsky, N., & Compton, R. (2005). Effects of practice, age, and task demands, on interference from a phone task while driving. *Accident Analysis & Prevention*, 37(2), 315–326.

Simon, H. A. (1957). *Models of Man: Social and National*. Wiley, New York, NY.

Simon, F., & Corbett, C. (1996). Road traffic offending, stress, age, and accident history among male and female drivers. *Ergonomics*, 39(5), 757–780.

Shorrock, S. T., & Williams, C. A. (2016). Human factors and ergonomics methods in practice: Three fundamental constraints. *Theoretical Issues in Ergonomics Science*, 17(5–6), 468–482.

Smithson, J. (2000). Using and analysing focus groups: Limitations and possibilities. *International Journal of Social Research Methodology*, 3(2), 103–119.

Society of Automotive Engineers (2002). Calculation of the time to complete in-vehicle navigation and route guidance tasks (SAE recommended practice J2365). Society of Automotive Engineers, Pennsylvania, PA.

Sodhi, M., Reimer, B., & Llamazares, I. (2002). Glance analysis of driver eye movements to evaluate distraction. *Behaviour Research Methods*, 34(4), 529–538.

Speake, J. (2015). 'I've got my Sat Nav, it's alright': Users' attitudes towards, and engagements with, technologies of navigation. *The Cartographic Journal*, 52(4), 345–355.

Srinivasan, R., & Jovanis, P. P. (1997). Effect of selected in-vehicle route guidance systems on driver reaction times. *Human Factors: The Journal of the Human Factors and Ergonomics Society*, 39(2), 200–215.

Stanton, N. A. (2015). Responses to autonomous vehicles. *Ingenia*, 62, 8.

Stanton, N. A., & Harvey, C. (2017). Beyond human error taxonomies in assessment of risk in sociotechnical systems: A new paradigm with the EAST 'broken-links' approach. *Ergonomics*, 60(2), 221–233.

Stanton, N. A., & Marsden, P. (1996). From fly-by-wire to drive-by-wire: Safety implications of automation in vehicles. *Safety Science*, 24(1), 35–49.

Stanton, N. A., & Salmon, P. M. (2009). Human error taxonomies applied to driving: A generic driver error taxonomy and its implications for intelligent transport systems. *Safety Science*, 47(2), 227–237.

Stanton, N. A., Salmon, P. M., Rafferty, L. A., Walker, G. H., Baber, C. R., & Jenkins, D.P. (2013). *Human Factors Methods: A Practical Guide for Engineering and Design*. Ashgate Publishing, Ltd., Aldershot, UK.

Stanton, N. A., Salmon, P. M., Walker, G. H., Salas, E., & Hancock, P. A. (2017). State-of-science: Situation awareness in individuals, teams and systems. *Ergonomics*, 60(4), 449–466.

Stanton, N. A., & Walker, G. H. (2011). Exploring the psychological factors involved in the Ladbroke Grove rail accident. *Accident Analysis & Prevention*, 43(3), 1117–1127.

Stanton, N. A., & Young, M. S. (1999). What price ergonomics? *Nature*, 399, 197–198.

Stanton, N. A., & Young, M. (2005). Driver behaviour with adaptive cruise control. *Ergonomics*, 48, 1294–1313.

Stanton, N. A., Young, M., & McCaulder, B. (1997). Drive-by-wire: The case of mental workload and the ability of the driver to reclaim control. *Safety Science*, 27(2–3), 149–159.

Stevens, A., & Minton, R. (2001). In-vehicle distraction and fatal accidents in England and Wales. *Accident Analysis & Prevention*, 33(4), 539–545.

Strauss, A., & Corbin, J. (1994). Grounded theory methodology. In Denzin, N. K., & Lincoln, Y. S. (Eds.) *The Sage Handbook of Qualitative Research*. Sage, Thousand Oaks, CA, 273–285.

Strayer, D. L., & Drews, F. A. (2004). Profiles in driver distraction: Effects of cell phone conversations on younger and older drivers. *Human Factors: The Journal of the Human Factors and Ergonomics Society*, 46(4), 640–649.

Strayer, D. L., Drews, F. A., & Johnston, W. A. (2003). Cell phone-induced failures of visual attention during simulated driving. *Journal of Experimental Psychology: Applied*, 9(1), 23.

Strayer, D. L., & Johnston, W. A. (2001). Driven to distraction: Dual-task studies of simulated driving and conversing on a cellular telephone. *Psychological Science*, 12(6), 462–466.

Stutts, J. C., Reinfurt, D. W., Staplin, L., & Rodgman, E. A. (2001). *The Role of Driver Distraction in Traffic Crashes*. AAA Foundation for Traffic Safety, Washington, DC.

Sullman, M. J. (2012). An observational study of driver distraction in England. *Transportation Research Part F: Traffic Psychology and Behaviour*, 15(3), 272–278.

Summala, H., Lamble, D., & Laakso, M. (1998). Driving experience and perception of the lead car's braking when looking at in-car targets. *Accident Analysis & Prevention*, 30(4), 401–407.

Summala, H. (1996). Accident risk and driver behaviour. *Safety Science*, 22(1–3), 103–117.

Summala, H. (2007). Towards understanding motivational and emotional factors in driver behaviour: Comfort through satisficing. In *Modelling Driver Behaviour in Automotive Environments*. Springer, London, UK. 189–207.

Terry, H. R., Charlton, S. G., & Perrone, J. A. (2008). The role of looming and attention capture in drivers' braking responses. *Accident Analysis & Prevention*, 40(4), 1375–1382.

The Highway Code. www.gov.uk/guidance/the-highway-code. Department for Transport. Updated 29 March 2016. Accessed 21/07/2016.

Theeuwes, J. (2004). Top-down search strategies cannot override attentional capture. *Psychonomic Bulletin & Review*, 11(1), 65–70.

THINK! (2017). Mobile Phones. http://think.direct.gov.uk/mobile-phones.html. Accessed 20/05/2017.

Tijerina, L., Parmer, E., & Goodman, M. J. (1998). Driver workload assessment of route guidance system destination entry while driving: A test track study. In *Proceedings of the 5th ITS World Congress,*12–16, Seoul, South Korea.

Tingvall, C., Ekstein, L., & Hammer, M. (2009) Government and industry perspectives on driver distraction. In Regan, M. A., Lee, J. D., Young, K. L. (Eds.) *Driver Distraction: Theory, Effects and Mitigation*. CRC Press, Boca Raton, FL.

References

Tingvall, C., & Haworth, N. (2000). Vision zero: An ethical approach to safety and mobility. In *6th ITE International Conference Road Safety & Traffic Enforcement: Beyond* 2000 September, Melbourne, VIC.

Tison, J., Chaudhary, N., & Cosgrove, L. (2011). National phone survey on distracted driving attitudes and behaviors. *Report No: DOT HS 811 555*, National Highway Traffic Safety Administration, Washington, DC.

Tivesten, E., & Dozza, M. (2015). Driving context influences drivers' decision to engage in visual-manual phone tasks: Evidence from a naturalistic driving study. *Journal of Safety Research*, 53, 87–96.

Tonetto, L. M., & Desmet, P. M. (2016). Why we love or hate our cars: A qualitative approach to the development of a quantitative user experience survey. *Applied Ergonomics*, 56, 68–74.

Törnros, J. E., & Bolling, A. K. (2005). Mobile phone use—Effects of handheld and hands-free phones on driving performance. *Accident Analysis & Prevention*, 37(5), 902–909.

Tractinsky, N., Ram, E. S., & Shinar, D. (2013). To call or not to call—That is the question (while driving). *Accident Analysis & Prevention*, 56, 59–70.

Trafikförordning (1998). (Traffic Regulation) §SFS. Rättsnätet Notisum AB. Chapter 4. 10e, 1276.

Transport Select Committee (2016). *Road Traffic Law Enforcement*. The Stationary Office Limited, London, UK.

Treat, J. R. (1980). A study of precrash factors involved in traffic accidents. *The HSRI Research Review*, 10(6), 35.

Treffner, P. J., & Barrett, R. (2004). Hands-free mobile phone speech while driving degrades coordination and control. *Transportation Research Part F: Traffic Psychology and Behaviour*, 7, 229–246.

Trick, L. M., Enns, J. T., Mills, J., & Vavrik, J. (2004). Paying attention behind the wheel: A framework for studying the role of attention in driving. *Theoretical Issues in Ergonomics Science*, 5(5), 385–424.

Trist, E. & Emery, F. (2005). Sociotechnical systems theory. In Miner, J. B. (Ed.) *Organizational Behavior 2: Essential Theories of Process and Structure*. ME Sharpe, Armonk, NY.

Trotter, M. J., Salmon, P. M., & Lenne, M. G. (2014). Impromaps: Applying Rasmussen's risk management framework to improvisation incidents. *Safety Science*, 64, 60–70.

Tsimhoni, O., Smith, D., & Green, P. (2004). Address entry while driving: Speech recognition versus a touch-screen keyboard. *Human Factors: The Journal of the Human Factors and Ergonomics Society*, 46, 600–610.

Underwood, G., Crundall, D., & Chapman, P. (2011). Driving simulator validation with hazard perception. *Transportation Research Part F: Traffic Psychology and Behaviour*, 14(6), 435–446.

Underwood, P., & Waterson, P. (2013). Systemic accident analysis: Examining the gap between research and practice. *Accident Analysis & Prevention*, 55, 154–164.

Vicente, K. J., & Christoffersen, K. (2006). The Walkerton E. coli outbreak: A test of Rasmussen's framework for risk management in a dynamic society. *Theoretical Issues in Ergonomics Science*, 7(02), 93–112.

Victor, T. W., Engström, J., & Harbluk, J. L. (2008). Distraction assessment methods based on visual behavior and event detection. In Regan, M. A., Lee, J. D., & Young, K. (Eds.) *Driver Distraction: Theory, Effects, and Mitigation*. CRC Press, Boca Raton, FL, 135–168.

Violanti, J. (1998) Cellular phones and fatal traffic collisions. *Accident Analysis and Prevention*, 30, 519–524.

Von Bertalanffy, L. (1968). *General System Theory*. George Braziller Inc., New York.

Vrkljan, B. H., & Polgar, J. M. (2007). Driving, navigation, and vehicular technology: Experiences of older drivers and their co-pilots. *Traffic Injury Prevention*, 8(4), 403–410.

Waddell, L. P., & Wiener, K. K. (2014). What's driving illegal mobile phone use? Psychosocial influences on drivers' intentions to use hand-held mobile phones. *Transportation Research Part F: Traffic Psychology and Behaviour*, 22, 1–11.

Walker, G. H., Gibson, H., Stanton, N. A., Baber, C., Salmon, P., & Green, D. (2007). Event analysis of systemic teamwork (EAST): A novel integration of ergonomics methods to analyse C4i activity. *Ergonomics*, 49(12–13), 1345–1369.

Walker, G. H., Stanton, N. A., & Chowdhury, I. (2013). Self-Explaining Roads and situation awareness. *Safety Science*, 56, 18–28.

Walker, G. H., Stanton, N. A., & Salmon, P. (2015). *Human Factors in Automotive Engineering and Design*. Human Factors in Transport Series. Ashgate Publishing Ltd., Aldershot.

Walker, G. H., Stanton, N. A., Salmon, P. M., & Jenkins, D. P. (2008). A review of socio-technical systems theory: A classic concept for new command and control paradigms. *Theoretical Issues in Ergonomics Science*, 9(6), 479–499.

Walker, G. H., Stanton, N. A., & Young, M. S. (2001). Where is computing driving cars? *International Journal of Human-Computer Interaction*, 13(2), 203–229.

Walsh, S. P., White, K. M., Hyde, M. K. & Watson, B. C. (2008) Dialling and driving: Factors influencing intentions to use a mobile phone while driving. *Accident Analysis and Prevention*, 40(6) 1893–1900.

Wang, Y., Mehler, B., Reimer, B., Lammers, V., D'Ambrosio, L. A., & Coughlin, J. F. (2010). The validity of driving simulation for assessing differences between in-vehicle informational interfaces: A comparison with field testing. *Ergonomics*, 53(3), 404–420.

Wang, J. S., Knipling, R. R., & Goodman, M. J. (1996). The role of driver inattention in crashes: New statistics from the 1995 Crashworthiness Data System. In *40th Annual Proceedings of the Association for the Advancement of Automotive Medicine,* 377, 392, Chicago, IL.

Ward, N. J. (2000). Automation of task processes: An example of intelligent transportation systems. *Human Factors and Ergonomics in Manufacturing & Service Industries*, 10(4), 395–408.

Weinberger, O. (1999). Legal validity, acceptance of law, legitimacy: Some critical comments and constructive proposals. *Ratio Juris*, 12(4), 336–353.

Welsh, E. (2002). Dealing with data: Using NVivo in the qualitative data analysis process. In *Forum Qualitative Sozialforschung (Forum: Qualitative Social Research),* 3(2), 1–9.

West, R., French, D., Kemp, R., & Elander, J. (1993). Direct observation of driving, self reports of driver behaviour, and accident involvement. *Ergonomics*, 36(5), 557–567.

White, M. P., Eiser, J. R., & Harris, P. R. (2004). Risk perceptions of mobile phone use while driving. *Risk Analysis*, 24(2), 323–334.

Wickens, C. D. (2002). Multiple resources and performance prediction. *Theoretical Issues in Ergonomics Science*, 3(2), 159–177.

Wierwille, W. W. (1993). Demands on driver resources associated with introducing advanced technology into the vehicle. *Transportation Research Part C: Emerging Technologies*, 1(2), 133–142.

Wierwille, W. W., Hanowski, R. J., Hankey, J. M., Kieliszewski, C. A., Lee, S. E., Medina, A., Keisler, A. S., & Dingus, T. A. (2002). Identification and evaluation of driver errors: overview and recommendations. *Report No: FHWA-RD-02-003*, US Department of Transportation, Federal Highway Administration, Washington, DC.

Wogalter, M. S., & Mayhorn, C. B. (2005). Perceptions of driver distraction by cellular phone users and nonusers. *Human Factors: The Journal of the Human Factors and Ergonomics Society*, 47(2), 455–467.

World Health Organisation (2004). World Report on Road Traffic Injury Prevention. World Health Organisation Report. http://apps.who.int/iris/bitstream/10665/42871/1/9241562609.pdf. Accessed 01/08/2017.

World Health Organisation (2011). *Mobile Phone Use: A Growing Problem of Driver Distraction*. WHO, Geneva, Switzerland.

References

World Health Organisation (2013). *WHO Global Status Report on Road Safety 2013: Supporting a Decade of Action*. WHO, Geneva, Switzerland.

World Health Organisation (2016). Road traffic Injuries Fact Sheet. www.who.int/mediacentre/factsheets/fs358/en/Reviewed. Accessed 28/10/2016.

Young, K. L., & Lenné, M. G. (2010). Driver engagement in distracting activities and the strategies used to minimise risk. *Safety Science*, 48(3), 326–332.

Young, K. L., Mitsopoulos-Rubens, E., Rudin-Brown, C. M., & Lenné, M. G. (2012). The effects of using a portable music player on simulated driving performance and task-sharing strategies. *Applied Ergonomics*, 43(4), 738–746.

Young, K. L., & Regan, M. (2007). Driver distraction: A review of the literature. In: Faulks, I. J., Regan, M., Stevenson, M., Brown, J., Porter, A., & Irwin, J. D. (Eds.) *Distracted Driving*. Australasian College of Road Safety, Sydney, NSW, 379–405.

Young, K. L, Regan, M., & Lee, J. D. (2008a). Measuring the effects of driver distraction: Direct driving performance methods and measures. In Regan, M. A., Lee, J. D., & Young, K. L. (Eds.), *Driver Distraction: Theory, Effects and Mitigation*. CRC Press, Boca Raton, FL, 85–107.

Young, K. L., Regan, M. A., & Lee, J. D. (2008b). Factors moderating the impact of distraction on driving performance and safety. In Young, K., Lee, J. D., & Regan, M. A. (Eds.) *Driver Distraction: Theory, Effects, and Mitigation*. CRC Press, Boca Raton, FL, 335–353.

Young, K. L., Rudin-Brown, C. M., Lenné, M. G., & Williamson, A. R. (2012). The implications of cross-regional differences for the design of In-vehicle Information Systems: A comparison of Australian and Chinese drivers. *Applied Ergonomics*, 43(3), 564–573.

Young, K. L., & Salmon, P. M. (2012). Examining the relationship between driver distraction and driving errors: A discussion of theory, studies and methods. *Safety Science*, 50(2), 165–174.

Young, K. L., & Salmon, P. M. (2015). Sharing the responsibility for driver distraction across road transport systems: A systems approach to the management of distracted driving. *Accident Analysis & Prevention*, 74, 350–355.

Young, K. L., Salmon, P. M., & Cornelissen, M. (2013a). Distraction-induced driving error: An on-road examination of the errors made by distracted and undistracted drivers. *Accident Analysis & Prevention*, 58, 218–225.

Young, K. L., Salmon, P. M., & Cornelissen, M. (2013b). Missing links? The effects of distraction on driver situation awareness. *Safety Science*, 56, 36–43.

Young, M. S., Birrell, S. A., & Stanton, N. A. (2011). Safe driving in a green world: A review of driver performance benchmarks and technologies to support 'smart' driving. *Applied Ergonomics*, 42(4), 533–539.

Young, M. S., & Lenné, M. G., (2017). *Simulators for Transportation Human Factors: Research and Practise*. CRC Press, Boca Raton, FL.

Young, M. S., Mahfoud, J. M., Stanton, N. A., Salmon, P. M., Jenkins, D. P., & Walker, G. H. (2009). Conflicts of interest: The implications of roadside advertising for driver attention. *Transportation Research Part F: Traffic Psychology and Behaviour*, 12(5), 381–388.

Young, M. S., & Stanton, N. A. (2002a). Malleable attentional resources theory: A new explanation for the effects of mental underload on performance. *Human Factors: The Journal of the Human Factors and Ergonomics Society*, 44(3), 365–375.

Young, M. S., & Stanton, N. A. (2002b). Attention and automation: New perspectives on mental underload and performance. *Theoretical Issues in Ergonomics Science*, 3(2) 178–194.

Young, M. S., & Stanton, N. A. (2004). Taking the load off: Investigations of how Adaptive Cruise Control affects mental workload. *Ergonomics*, 47(8), 1014–1035.

Young, M. S., & Stanton, N. A. (2007). Miles away: Determining the extent of secondary task interference on simulated driving. *Theoretical Issues in Ergonomics Science*, 8(3), 233–253.

Zhou, R., Rau, P. L. P., Zhang, W., & Zhuang, D. (2012). Mobile phone use while driving: Predicting drivers' answering intentions and compensatory decisions. *Safety Science*, 50(1), 138–149.

Zhou, R., Wu, C., Rau, P. L. P., & Zhang, W. (2009). Young driving learners' intention to use a handheld or hands-free mobile phone when driving. *Transportation Research Part F: Traffic Psychology and Behaviour*, 12(3), 208–217.

Author Index

A

Ajzen, I., 32, 90
Alliance of Automobile Manufacturers, 15–16, 187, 200–1
Allison, C. K., 156
Alm, H., 20, 26, 34, 80, 121, 131, 148, 155, 175, 192
Atchley, P., 114, 133, 137
Axon, S., 132

B

Baber, C., 7
Banks, V. A., 60, 153, 202
Bayliss, D., 93
Beanland, V., 12–13, 38, 89, 102, 114, 147, 149, 176, 195
Bella, F., 175
Bittner, A., 174
Blomquist, G., 77
Boer, E. R., 148
Bojko, A. A., 128
Boyatzis, R. E., 96–7
Branford, K., 35, 84, 115
Braun, V., 95–6, 103, 115
Brodsky, W., 8,11
Brookhuis, K. A., 50, 148, 201
Brown, I. D., 1
Bunn, T. L., 3
Burnett, G. E., 1–2, 10, 12, 20, 30, 42, 57–8, 61, 77, 90, 132, 149, 182
Burns, P. C., 174
Byrne, E. A., 202

C

Cacciabue, P. C., 7
Caird, J. K., 10–11
Campbell, J. L., 97
Campbell, S. W., 203
Cantin, V., 174
Carsten, O., 11, 30, 33, 113, 149, 174, 193
Cassano-Piche, A. L., 45, 65
Cavaye, A. L., 47
Charlton, S. G., 139
Chen, H. Y. W., 32, 64, 121, 192
Cnossen, F., 8, 42, 44, 47, 49–50, 53, 57, 61, 112–13, 122, 133, 148
Cohen D., 93

Commission of the European Communities, 16, 71, 187, 200
Cooper, P. J., 27
Cummings, P., 11

D

Darke, P., 55
Davidsson, S., 118
de Waard, D., 203
Dekker, S.W., 2. 11. 13. 16–17, 20, 42, 64, 190, 198–9
Denzin, N. K., 21, 195
Department for Transport, 14, 31–2, 63–4, 66–7, 84, 93, 117, 144, 203
Dingus, T. A., 12–13, 31, 43, 90
Dixon, S. M., 123
Donmez, B., 42, 64, 192
Dogan, E., 42. 51. 112
Dong, Y., 10, 38, 58, 189, 191
Dorf, R. C., 11
Drews, F. A., 9, 42, 121–2, 142, 179
Driver and Vehicle Licensing Agency, 14
Dukic, T., 20
Dulisse, B., 75

E

Edwards, W., 147
Elvik, R., 34
Eost, C., 80
Ercikan, K., 20, 21, 195
Ericsson, K. A., 154, 174
Eriksson, A., 20
EURO NCAP, 38, 183, 188–9, 191, 201, 203

F

Fagnant, D. J., 189, 201
Fraedrich, E., 202
Fuller, R., 434, 53–4, 59
Furui, S., 136
Funkhouser, D., 22

G

Gardner, B., 93
Gardner, M., 133
Gillham, B., 54
Glaser, B.G., 46, 180

Godley, S. T., 149–50
Gonçalves, J., 10, 38
Goodman, M. J., 31
Gordon, C.P., 1
Green, P., 15, 39, 58, 121, 131, 192, 195
Groeger, J. A., 47

H

Haigney, D. E., 53, 148, 150
Hancock, P. A., 55, 79
Hancox, G., 29
Hanowski, R. J., 3, 10–11
Harms, L., 25, 42
Harbluk, J. L., 8, 23, 30
Hart, S. G., 156
Harvey, C., 2, 7, 15–16, 20, 35, 42–3, 58–9, 61,
 90, 93, 118, 141, 148, 172–3, 190,
 195–7, 199, 204
He, J., 20, 76
Healey, J. A., 23
Health and Safety at Work etc. Act 80
Hettinger, L. J., 16–18, 45–6, 189, 196, 199
Hickman, J. S., 3, 10–11
Hignett, S., 30, 103, 114, 177, 181, 183, 198
Hockey, G.R.J., 43–4, 50, 53–4, 61, 75, 148
Hoel, J., 1
Hollnagel, E., 17, 45
Holtzblatt, K., 150
Horberry, T., 119, 122, 139, 170, 175
Horrey, W. J., 2, 14, 34, 44, 50, 61, 64, 74, 79, 90,
 115, 142, 149, 150–1, 155, 171, 174, 187
Hosking, S. G., 49
Hedlund, J., 1, 8
Huemer, A. K., 91
Hughes, B. P., 19, 34, 38–9, 45

J

Jahn, G., 175
Janitzek, T., 2
Jamson, A. H., 42–3, 54, 189, 202
Japan Automobile Manufacturers Association,
 15, 187, 200
Jensen, B. S., 42, 57
Johansson, R., 34

K

Kahneman, D., 147, 182
Kaptein, N., 20, 82
Kazaras, K., 37
Kervick, A. A., 75
Kircher, K., 8, 33–4, 42–4, 51, 58, 77, 148, 174,
 198
Klauer, S. G., 11, 13, 15, 20, 31, 39, 74, 82, 90

Klein, G.A., 147–8, 150, 182
Konstantopoulos, P., 119
Korpinena, L., 203
Kountouriotis, G. K., 122
Kurasaki, K. S., 97
Kutila, M., 10

L

Lajunen, T., 119
Lansdown, T. C., 2, 17, 48, 64, 82, 90, 134, 149,
 172
Lam, L. T., 23, 195
Lamble, D., 34, 121–2, 137, 140, 142
Larsson, P., 2, 33–4, 58, 64, 82, 84, 117
Layder, D., 46
Lee, C., 136
Lee, J. D., 1–2, 8–9, 14, 20, 42–4, 50–1,
 60–1, 90, 92, 104, 114, 118, 149,
 179, 186
Lerner, N., 29, 32, 44, 86, 91, 102, 119, 121–2,
 133, 137, 139, 150, 172–3, 196
Leveson, N. G., 2, 12, 19, 33, 35, 37, 45, 51, 60,
 64, 68, 82, 103, 178, 181, 189, 197
Liang, Y., 10
Liu, Y.C., 191
Loukopoulos, L. D., 203

M

Manser, M. P., 1, 9, 179
Marshall, M. N., 123
Matthews, B. W., 157–9
McCartt, A. T., 12, 20, 31, 39, 42, 44, 115
McEvoy, S. P., 23, 34, 90, 119, 121–3, 140, 142,
 192
Mehlenbacher, B., 77
Meng, L., 132
Merat, N., 11, 122, 202
Metz, B., 28, 43, 53, 61, 90, 113, 119, 148, 150–1,
 171–2, 204
Michon, J. A., 148
Miller, D., 202
Mitsopoulos-Rubens, E., 42, 115
Mitzner, T. L., 121
Mizenko, A. J., 121, 133, 137
Monk, C. A., 50
Moray, N., 46
Musselwhite, C. B., 29, 32, 46

N

Naweed, A., 203
Neale, V. L., 12–13, 31, 90, 93
Nelson, E., 31, 118
Nemme, H. E., 133

Author Index

Newnam, S., 17, 65
NHTSA, 15–16, 19–20, 30–1, 39, 42, 57–8, 67, 74, 76, 141, 187, 190, 299, 201, 203
Noy, Y. I., 8, 23, 30, 44, 53, 112

O

O'Cathain, A., 32, 91
Office for National Statistics, 18

P

Papadakos, P. J., 223
Parnell, K. J., 17, 24, 64–5, 68, 79, 91, 102–3, 108–9, 112, 114–15, 131, 133, 143, 148, 150, 154–5, 157, 163, 171, 173, 177, 182–3, 186–7, 192–3, 198–9
Patel, J., 97, 106
Patten, C. J., 9, 25, 50, 174, 179
Patton, M. Q., 96–7
Perrow, C., 45
Pettitt, M.A., 1, 8, 42, 179
Phillips, J. K., 147, 182
Plant, K. L., 7, 54, 79, 97
Pope, C. N., 121–2, 140, 142
Postman, N., 16
Pöysti, L., 123, 137, 139–40, 142

R

RAC, 18, 93
Rafferty, L. A., 47, 49, 51, 54, 180
Rakauskas, M. E., 44, 50, 53, 122, 133, 148
Ranney, T. A., 8, 11, 22, 58, 149, 189, 199
Rasmussen, J., 21, 35–6, 45–6, 63, 65, 67–9, 79–80, 82, 87, 89, 115, 186, 194, 197–8, 205
Reason, J., 2, 7, 16, 57, 64, 198–9
Redelmeier, D. A., 2, 11, 24, 74, 187
Reed, M. P., 121, 131, 192,
Regan, M. A., 1, 8, 42–3, 47, 50, 76, 90, 179, 181, 186
Reimer, B., 20, 149
Reinach, S., 35–6
Richards, L., 95
Road Traffic Act, 74, 76
Ritchie J., 97
Road Safety Act 94
Robert, G., 50, 75
Rosencrantz, H., 34
Rudin-Brown, C. M., 119, 191, 200, 202

S

Sabey, B. E., 7
SAE 10, 202–3

Salmon, P. M., 1,2, 7–9, 17, 19, 20–1, 33, 35–6, 38–9, 43–6, 56, 61–2, 64–5, 68, 79, 82, 84, 90, 97, 101–2, 114–15, 117, 140, 142–3, 153, 158, 174, 185, 187, 189–90, 196–8, 203, 206
Salvucci, D., 50, 58
Sanders, M.S., 57
Sawyer, B. D., 2, 14, 64, 76
Scarborough, A., 35
Schömig, N., 28, 43, 53, 61, 90, 113, 119, 148, 150, 171–2
Scott-Parker, B., 17, 65
Schwebel, D. C., 203
Shappell, S. A., 19. 35–6, 68, 178, 181, 197
Sheridan, T. B., 1, 43
Shinar, D., 119
Simon, F., 93, 148, 154, 174
Shorrock, S. T., 19, 35
Smithson, J., 91, 119
Sodhi, M., 24
Speake, J., 131–2, 140
Stanton, N. A., 2, 7–8, 10, 14–17, 19–22, 33, 35, 39, 42–3, 46, 50, 54, 58–61, 69, 79, 82, 90, 97, 81, 114, 117, 140–1, 148, 157, 172–3, 189–90, 195–7, 202, 204
Stevens, A., 12, 38
Strauss, A., 46–7, 180
Strayer, D. L., 19, 24, 42, 53, 74, 121–2, 142, 149, 150, 175, 179, 187
Stutts, J.C., 2, 12–13, 31, 38, 90
Sullman, M. J., 2, 22, 31, 195
Summala, H., 20, 42, 119, 148–9

T

Terry, H. R., 2
The Highway Code., 64, 74, 76, 144
Theeuwes, J., 54, 61
THINK! 117
Tijerina, L., 15
Tingvall, C., 2, 35, 60, 62–3, 84, 187
Tison, J., 122, 137
Tivesten, E., 53, 90, 102, 113, 117–19, 147, 151, 171, 193
Tonetto, L. M., 93, 97
Törnros, J. E., 97
Trafikförordning, 64
Transport Select Committee, 71, 77
Treat, J.R., 1, 8
Treffner, P. J., 142
Trick, L. M., 1, 50
Trist, E., 17
Trotter, M. J., 65, 68, 82
Tsimhoni, O., 2, 14, 20, 61, 64, 115, 149

U

Underwood, G., 150
Underwood, P., 16, 19, 35, 196

V

Vicente, K. J., 65
Victor, T. W., 8
Von Bertalanffy, L., 2, 17
Vrkljan, B. H., 46

W

Waddell, L. P., 133
Walker, G. H., 1–2, 7, 17, 35, 54, 64, 67, 76, 79, 92–3, 107, 122, 140, 150, 153, 158, 192, 199
Walsh, S.P., 32, 90, 102, 114–15, 142, 145
Wang, J. S., 12

Welsh, E., 95
White, M. P., 29, 133, 196
Wickens, C. D., 2, 8, 14, 34, 43, 50–4, 58, 61, 64, 74, 112, 115, 142, 155, 174, 187
Wierwille, W. W., 7, 51
Wogalter, M. S., 79
World Health Organisation, 1–2, 63–4, 76, 189

Y

Young, K. L., 1–2, 8–9, 17, 19–20, 25, 30–3, 35–6, 38–9, 42–7, 51, 56, 61–2, 64–5, 74, 76–8, 84, 86, 90–3, 102, 114–15, 119, 142–3, 145, 153, 172, 179, 181, 185, 187, 191, 196–8, 200,
Young, M. S., 20, 22, 50, 149, 157, 182, 202

Z

Zhou, R., 32, 90, 102, 114–15, 145

Subject Index

A

AcciMap analysis 35, **70–84**, 178, 181
  analysis 69–77
    phone use 70–5
    other technology use 73, 75–7
    phone/other technology comparison
      77–9
  background 35
  example 36
  methodology 65–8
  phoney use case study 79–80, 81
  sat-nav case study 80–1, 83, *see also* Risk
    Management Framework (RMF)
Actor map 69–71, 188, 198
Adapt to demands
  case study 59
  definition 49–50, 69
  other road users theme 183–4
  PARRC interconnections 53, 110–12,
    184–5
  thematic framework 106, *see also* PARRC
    model
Age effects 91, 121–2, 140–4, 192
  likelihood ratings 127–30
    young 128–9
    middle 129–30
    older 130
Automated driving **201–3**
  adaptive cruise control (ACC) 202

B

Behavioural Regulation
  case study 60
  definition 50, 69
  PARRC interconnections 54, 110–14, 184
  thematic framework 106, *see also* PARRC
    model

C

Car following task 26, 30
Case study examples
  change destination 80, 82–3
  delivery driver 79–81
  Victoria McClure 41–2, 55–60
Chi-square test 160, 178
Complex system 17, 33, *see also* Sociotechnical
  System Theory

Context 16–18, 33, 39, 67, 100–2, 116, 118–19,
  150, 162, 164–5, 201, 214–15, 217
Control levels
  operational control 44, 148
  strategic control 44, 148–51, 173–4
  tactile control 44
Countermeasures 11, 17, 33, 35–6, 61, 85–6,
  115–16, 186–93, 205
Crash statistics 12–14
Cultural differences 119, 191, 200–1

D

Deductive analysis 96, 181, 183, *see also*
  Thematic analysis
Design guidelines 15, 12, 14, **15–16**, 67
  european guidelines 16, 71, 187, 200–1
  JAMA guidelines 15–16, 187, 200
  NHTSA guidelines 15–16, 19, 20, 30, 39, 57,
    67, 74, 75–6, 141, 187, 190, 199, 201,
    203
Driver distraction
  causal factors 49–51 (*see also* PARRC model)
  challenges 17,19–21, 202–3
  definition 1, 8–9, 179
  methods 21–32 (*see also* Methods
    dichotomies)
  sociotechnical systems definition 185–6, 200
  voluntary distraction 1, 89–92, 101–2, 114,
    147, 149
Driver error 7–8, 19, 33, 35, 84, 101, *see also*
  Human Error
Driver monitoring 10, 38, 189, 191
Driver performance metrics 24, **30**, 33, 174, 204
Driving simulator 11, 16, 22, 33, 113, 150–3,
  160–3, 171 174–6, 179, 182–4, 204,
  199
  advantages 183
  simulation 156
  simulator fidelity 20, 149
  simulator validity 20, 149–50, 171, *see also*
    Southampton University Driving
    Simulator (SUDS)

E

End-users 74, 80, 85–6, 118, 189, 187
Engagement regulation 99, 118, 165, 172–3, 191,
  214, *see also* PARRC model
Error taxonomies 7–8, 36, 68, 84, 97, 101, 140

247

EURO NCAP 38, 189, 191, 201, 203
Experimental environment 149–50, 204–5,
    see also Driving Simulators and
    On-road study
Eye tracking 19, 21, 24, 58, 175–204
Eyes-off-the-road time 15–16, 31, 55, 118

F

Focus groups 29, 32–3, 86, 91, 119
Fatigue 2–3, 11, 48, 58
Feedback
    bottom-up 45, 54, 56, 67, 87, 96, 113,
        114, 118
    loops 45, 67
    top-down 54, 57–8, 67, 87, 117, 183, 187

G

Gap acceptance task 27, 30
Gender effects 122, 125, 140, 144–5, 192
Goal-directed behaviour 42, 47
    goal conflict
        case study 57
        definition 50–1, 69
        PARRC interconnections 51–4, 110–13,
            114, 185, 197
        thematic framework 106–7
    Goal Priority
        case study 59
        definition 51, 69
        PARRC interconnections 51–4, 110–12,
            173, 184–5, 200
        thematic framework 107, *see also PARRC
            model*
Government 14–15, 45, 56, 57, 58, 61, 65, 66–7,
        70–1, 84–5, 114–15, 187–8, 206
Grounded theory **46–7**, 51, 54, 178, 180–1
    advantages 47
    document analysis 47
    exclusion criteria 47–8
    inclusion criteria 48
    search criteria 47

H

Hand-held phone use
    AcciMap analysis 71–5
    likelihood ratings/reasoning 137–9, 172–3
    legislation 2, 14, 64 (*see* Legislation)
    protective effects 10–11, *see also* Text
        messaging
Hands-free phone use
    legislation 142 (*see* Legislation)
    likelihood ratings/reasoning 132–3
Health and Safety at Work Act 80
Heat mapping 127–31

Hindsight Bias 8, 16–17, 33, 43, 92, 198–9
Human Factors Analysis and Classification
        System (HFACS) 35–6, 38, 68, 82,
        84, 181
Human error 16–17, 34, 196
Human Machine Interface (HMI) 57, 57, 74,
        76–7, 85, 101, 187, 191, 192
100-car study 13, 31
    crash statistics 13
    eyes-off-road time 15

I

In-vehicle technologyuse 1, 9–12, 65–8, *see*
        Handheld phone, Hands-free phone,
        IVIS, Satnav, Voice command system
Inattention 1, **8**, 43
Individualistic approach 2–3, 9, 16–17, 34, 62,
        187, 191–2, 198
Inductive analysis 92, **96–7**, 98, 115, 120, 181–2
    descriptive themes 98, 117, 201–15
    semantic themes 96–7, 98–100, 104–5, 183–4
    systemic themes 98–100, 115, 117–18,
        219–20, *see also* Thematic analysis
Industrialists 66–7, 85, 116, 189–90, 205
Infrastructure 59, 99–102, **117**, 139, 162, 165–7,
        173–4, 187, 192–3, 199, 209–10,
        219–20
Instrumented vehicle (IV) 151–3
International committees 65–6, 70–1, 85,
        115–16, 187–8, 205, *see also* Risk
        management framework
In-vehicle infotainment system (IVIS) 108, 140,
        141–2, 145, 155, 191
    likelihood ratings/reasoning 134–5

J

Junctions 95, 129–30, 137, 139, 144, 173, 192–3,
        201, 220, *see also* Roundaboutsand
        Road type

L

Lane-change task (LCT) 25
Legislation 2, 5, 12, 14–15, 56, 58–9, 61–2,
        63–5, 85, 87, 140, 142, 187–8, 192,
        205
    hand-held phone use 2, 14, 64, 70–5
    other technology use 12, 15, 61, 73, 75–7, 142
Likert scale 21, 93, 123, 125
    ratings 127–30
    reasoning 130–9
Line of appropriate behaviour 79, 80, *see also*
        Sociotechnical Systems Theory
Lobbyists 66–7, *see also* Risk Management
        Framework

Subject Index

M

Matrix query 103, 108–9, 112, 125, 131, 145, **163–7**, 183, *see also* Nvivo
Matthews correlation coefficient (Phi) 157, 159, 204
Media campaigns 31–2, 59, 61, 69–70, 74, 77, 85–6, 188, 192, 199
Methods dichotomies 30, 195
  classification 22
  objective qualitative 31
  objective quantitative 30
  subjective qualitative 32–2
  subjective quantitative 31–2
Minimum required Attention theory (MiRA) 42, 58, 148
Mobile phone manufacturers 143–4, 173, 189–90, 205
Multiple Resource Theory (MRT) 8, 43, 52

N

NASA-TLX 156, *see also* Workload
National committees 66, 71, 76, 85, 187, 197, 203, *see also* Risk Management Framework
Naturalistic Decision Making (NDM) 147, 182, 201
  verbal protocol method 174
Naturalistic driving studies 13, 31, 102, 113, **149–51**
Nvivo 95, 97, 103–4, 125, 131, 145, 157, 163

O

Observation studies 22, 32–3
Occlusion task 15–16, 19, 199
On-road study 23, 30
  data analysis 157–8
  participants 151–2
  procedure 153, *see also* Naturalistic driving studies
Online survey 86, 123–7, 203
  data analysis 125
  participants 124
  procedure 125
Other road users 75, 102, 117, 161–3, 171, 173, 175, 183–4, 219–20

P

PARRC model
  application
    AcciMap analysis 69–77
    case study 54–60
    thematic framework 106–13, 182–4
  development 54–60, 180

evolution 178–9
  factors **49–51**
  interconnections
    driving study 182–5
    grounded theory 51–4
    interviews 108–13
  validation 113–14, **177**, 181–5
Percentage agreement 97–8, *see also* Rater reliability
Peripheral detection task (PDT) 25, 30
Physiological measures 23
Pilot study 95, 156, *see also* Semi-structured Interviews
Public opinion 67, 84, 89

Q

Qualitative methods 21, 22–5, 29, 31, 32–3
Quantitative methods 21, 22–9, 30, 31

R

Regulators 66–7, 70, 74, 85, 115–16, **188–9**
Rater Reliability 97–8
Research-practise gap 4, 6, 19, 186, 194, 196
Resource constraints
  case study 58–9
  definition 51, 69
  PARRC interconnections 51–4, 110–14, 173, 184–5, 200
  thematic framework 107–8, *see also* PARRC model
Responsibility 9, 17, 21, 39, 46, 55, 60, 61, 66, 74–5, 80, 87, 143–4, 199, 205
Risk Management Framework (RMF) 63, **65–7**, 68–70, 72, 82, 84–6, 115–16, 186–8, 197–8, 201, 205
Road type 90, 93, 117–18, 122, 123, 127–8, 139, 158, 160–1, 163, 167, 171–2, 173, 201
  junctions 93, 95, 100, 129, 134, 173–4, 201
  motorway 100, 106, 117,130, 132–3, 138, 39, 156, 158, 160–1, 165–70, 173, 175, 219–20
  residential 93, 132
  rural 93, 117, 133
  urban 93, 117, 133
Roundabouts 152, 156, 158, 160, 161, 163, 165–7, 173–4, 201, 219–20, *see also* Junctions

S

SAE 10, 202, 203
Satellite navigation systems (Satnav) 2, 20, 82, 91, 131, 101, 117, 190, 192
  case study 41–2, 54–60, 80–2
  legislation 77 (*see* Legislation)

Satellite navigation systems (Satnav) *(cont.)*
likelihood ratings/reasoning 95, 128, 130, 131–2, 140–1, 145, 155, 160, 165, 172
Satisficing 148
Scree plot 48–9
Self-report measures 29, 31–2, 90, 92, 98, 101, 119, 122, 144
Semi-structured interviews 29, 32, 91, **123**, 181
advantages 32, 119
data analysis 95–7, 103–4
participants 92–3, 123–4
procedure 94–4, 124
SHRP 2, 13
Situational awareness 17, 25, 93, 122, 139–40, 148, 174, 202
Sociotechnical System Theory 2–4, **16–18**, 38, 200, 206
accident analysis 16–18
methodology 33–9
Southampton University Driving Simulator (SUDS) 152, *see also* Driving Simulator
Speed metrics 168–70
Systems Theoretic Accident Model and Process (STAMP) 35, 37–8, 68, 82, 84, 181
Systems pressures 67, 80, 133, 190

T

Task-capability interface (TCI) mode 144, 53, 59
Taxonomies 7–8, 36, 68, 84, 97, 101, 140, 178
Technological revolution 9, 205

Test-track studies 22, 23, 25, 30, 90, 149–51, 171
Text message 21, 137, 143, 145, **217–18**
on-road study 155, 160, 164–5, 172–3
semi-structured interviews 94, 107, 114, *see also* Hand-held mobile phone
Thematic analysis 92, **95–8**, 103, 115–16, 178
thematic framework **98**, 105, 162, 207, 214, *see also* Inductive analysis
Theory of planned behaviour (TPB) 32, 90, 192
Triangulation 103, 177, 183, 186, 193
Truck drivers 10–11, 203

V

Verbal protocol 25, 34–5, 151–2, 174, 182–3, 204
methodology 153–4
Analysis 157–8, *see also* Naturalistic Decision Making
Video rating 29
Voice command system 85, 94, 132, 135–6, 141–2, 159, 164, 166, 189
likelihood ratings/reasoning 135–6

W

Willingness to engage 28–9, 31, 57, 77, 86, 91, 119, 133, 145
intention 32, 90, 102, 114–15, 150–1, 157–8, 159–63, 171, 183–4
Workload 58, 156, 158, 170, 174–5, 202, *see also* NASA-TLX
Workload managers 58, 192